Graphs, Surfaces and Homology
Third Edition

Homology theory is a powerful algebraic tool that is at the centre of current research in topology and its applications. This accessible textbook will appeal to students interested in the application of algebra to geometrical problems, specifically the study of surfaces (such as sphere, torus, Möbius band, Klein bottle). In this introduction to simplicial homology – the most easily digested version of homology theory – the author studies interesting geometrical problems, such as the structure of two-dimensional surfaces and the embedding of graphs in surfaces, using the minimum of algebraic machinery and including a version of Lefschetz duality.

Assuming very little mathematical knowledge, the book provides a complete account of the algebra needed (abelian groups and presentations), and the development of the material is always carefully explained with proofs of all the essential results given in full detail. Numerous examples and exercises are also included, making this an ideal text for undergraduate courses or for self-study.

PETER GIBLIN is a Professor of Mathematics (Emeritus) at the University of Liverpool.

Graphs, Surfaces and Homology
Third Edition

PETER GIBLIN
University of Liverpool

CAMBRIDGE
UNIVERSITY PRESS

University Printing House, Cambridge CB2 8BS, United Kingdom

One Liberty Plaza, 20th Floor, New York, NY 10006, USA

477 Williamstown Road, Port Melbourne, VIC 3207, Australia

314-321, 3rd Floor, Plot 3, Splendor Forum, Jasola District Centre, New Delhi - 110025, India

79 Anson Road, #06-04/06, Singapore 079906

Cambridge University Press is part of the University of Cambridge.

It furthers the University's mission by disseminating knowledge in the pursuit of education, learning and research at the highest international levels of excellence.

www.cambridge.org
Information on this title: www.cambridge.org/9780521154055

© P. Giblin 1977, 1981, 2010

This publication is in copyright. Subject to statutory exception and to the provisions of relevant collective licensing agreements, no reproduction of any part may take place without the written permission of Cambridge University Press.

First and second edition published by Chapman and Hall
Third edition published by Cambridge University Press

A catalogue record for this publication is available from the British Library

Library of Congress Cataloging in Publication data
Giblin, P. J.
Graphs, surfaces, and homology / Peter Giblin. – 3rd ed.
p. cm.
Includes bibliographical references and index.
ISBN 978-0-521-76665-4 (hardback) – ISBN 978-0-521-15405-5 (pbk.)
1. Algebraic topology. I. Title.
QA612.G5   2010
514´.2–dc22   2010022763

ISBN 978-0-521-76665-4 Hardback
ISBN 978-0-521-15405-5 Paperback

Cambridge University Press has no responsibility for the persistence or accuracy of URLs for external or third-party internet websites referred to in this publication, and does not guarantee that any content on such websites is, or will remain, accurate or appropriate.

For Rachel

Contents

Preface to the third edition	page	xi
Preface to the first edition		xiii
List of notation		xvii
Introduction		1
1 Graphs		**9**
Abstract graphs and realizations		9
★ Kirchhoff's laws		14
Maximal trees and the cyclomatic number		16
Chains and cycles on an oriented graph		20
★ Planar graphs		26
★ Appendix on Kirchhoff's equations		35
2 Closed surfaces		**38**
Closed surfaces and orientability		39
Polygonal representation of a closed surface		45
★ A note on realizations		47
Transformation of closed surfaces to standard form		49
Euler characteristics		55
★ Minimal triangulations		60
3 Simplicial complexes		**67**
Simplexes		67
Ordered simplexes and oriented simplexes		73
Simplicial complexes		74
Abstract simplicial complexes and realizations		77
Triangulations and diagrams of simplicial complexes		79
Stars, joins and links		84

	Collapsing	88
	★ Appendix on orientation	93
4	**Homology groups**	**99**
	Chain groups and boundary homomorphisms	99
	Homology groups	104
	Relative homology groups	112
	Three homomorphisms	121
	★ Appendix on chain complexes	124
5	**The question of invariance**	**127**
	Invariance under stellar subdivision	128
	★ Triangulations, simplicial approximation and topological invariance	133
	★ Appendix on barycentric subdivision	136
6	**Some general theorems**	**138**
	The homology sequence of a pair	138
	The excision theorem	142
	Collapsing revisited	144
	Homology groups of closed surfaces	149
	The Euler characteristic	154
7	**Two more general theorems**	**158**
	The Mayer–Vietoris sequence	158
	★ Homology sequence of a triple	167
8	**Homology modulo 2**	**171**
9	**Graphs in surfaces**	**180**
	Regular neighbourhoods	183
	Surfaces	187
	Lefschetz duality	191
	★ A three-dimensional situation	195
	Separating surfaces by graphs	198
	Representation of homology elements by simple closed polygons	200
	Orientation preserving and reversing loops	203
	A generalization of Euler's formula	207
	★ Brussels Sprouts	211
	Appendix: abelian groups	**215**
	Basic definitions	215
	Finitely generated (f.g.) and free abelian groups	217

Quotient groups	219
Exact sequences	221
Direct sums and splitting	222
Presentations	226
Rank of a f.g. abelian group	233
References	239
Index	243

Preface to the third edition

Since the second edition of this book, published in 1977, went out of print I have received what a less modest person might describe as fan-mail, lamenting the fact that a really accessible introduction to algebraic topology, through simplicial homology theory, was no longer available. Despite the lapse of years, and despite the multiplicity of excellent texts, nothing quite like this book has appeared to take its place. That is the reason why the book is being reprinted, newly and elegantly typeset, and with all the figures re-drawn, by Cambridge University Press. (I don't think TEX was even a gleam in the eye of that wonderful benefactor of mathematicians, Donald Knuth, when the first edition of the book was beautifully typed on a double keyboard mechanical typewriter by the then Miss Ann Garfield, now Mrs Ann Newstead.)

Of course, homology theory has advanced in the interim – not just the theory, but most importantly the multiplicity of applications and interactions with other areas of mathematics and other disciplines. Shape description, robotics, knot theory (Khovanov homology), algebraic geometry and theoretical physics – to name just a few areas – use topological ideas and in particular homology and cohomology. A quick search with an Internet search engine will turn up many references to applications. Since this book is intended for beginners, there is no pretence of being able to cover the recent developments and I have confined my 'updating' to correcting obvious errors, replacing references to other textbooks with more accessible modern ones, including additional references to the research literature and adding some comments where it seemed appropriate. The driving force behind the book remains to present a thoroughly accessible and self-contained introduction to homology theory via its most easily digested manifestation, simplicial homology.

I thank again those who helped with the first and second editions, namely, in alphabetical order, Ronnie Brown, Bill Bruce, Pierre Damphousse, Chris Gibson, Erwin Kronheimer, Hugh Morton, Ann Newstead, John Reeve, Stephen

Wilson and Shaun Wylie. I now thank Jon Woolf for his advice, and Cambridge University Press for taking on a new edition – and, of course, my students and those who wrote telling me that they liked my book.

A final word on literary manners, which change over the years. In re-reading the text I was conscious of the use of 'he' and 'his' when referring to 'the reader'. At the time of writing the first edition, everyone understood that these meant 'he or she' and 'his or her', but we have become more sensitive to gender issues in the meantime. Rather than laboriously change all these by some circumvention or other let me state here, once for all, that I truly hope to interest female and male readers without distinction, and I always use the pronouns, when not referring to a specific person, to mean either gender.

Preface to the first edition

Topology is pre-eminently the branch of mathematics in which other mathematical disciplines find fruitful application. In this book the algebraic theory of abelian groups is applied to the geometrical and topological study of objects in euclidean space, by means of homology theory. Several books on algebraic topology contain alternative accounts of homology theory; mine differs from these in several respects.

Firstly the book is intended as an undergraduate text and the only mathematical knowledge which is explicitly assumed is elementary linear algebra. In particular I do not assume, in the main logical stream of the book, any knowledge of point-set (or 'general') topology. (There are a number of tributaries, not part of the main stream but I hope no less logical, about which more in a moment.) A reader who is familiar with the concept of a continuous map will undoubtedly be in a better position to appreciate the significance of homology theory than one who is not but nevertheless the latter will not be at a disadvantage when it comes to understanding the proofs.

The avoidance of point-set topology naturally imposes certain limitations (in my view quite appropriate to a first course in homology theory) on the material which I can present. I cannot, for example, establish the topological invariance of homology groups. A weaker result, sufficient nevertheless for our purposes, is proved in Chapter 5, where the reader will also find some discussion of the need for a more powerful invariance theorem and a summary of the proof of such a theorem.

Secondly the emphasis in this book is on low-dimensional examples – the graphs and surfaces of the title – since it is there that geometrical intuition has its roots. The goal of the book is the investigation in Chapter 9 of the properties of graphs in surfaces; some of the problems studied there are mentioned briefly in the Introduction, which contains an informal survey of the material of the book. Many of the results of Chapter 9 do indeed generalize to higher dimensions (and the general machinery of simplicial homology theory is available from earlier chapters) but I have confined myself to one example, namely the theorem that non-orientable closed surfaces do not embed in three-dimensional space. One of the principal results of Chapter 9, a version of Lefschetz duality, certainly generalizes, but for an effective presentation such a generalization needs cohomology theory. Apart from a brief mention in connexion with Kirohhoff's laws for an electrical network I do not use any cohomology here.

Thirdly there are a number of digressions, whose purpose is rather to illuminate the central argument from a slight distance, than to contribute materially to its exposition. (To change the metaphor, these are the tributaries mentioned earlier. Some of them would turn into deep lakes if we were to pursue them far.) The longest digression concerns planar graphs, and is to be found in Chapter 1, but there are others on such topics as Kirchhoff's laws (also Chapter 1), minimal triangulations, embedding and colouring problems (Chapter 2), orientation of vector spaces (Chapter 3) and a game due to J. H. Conway (Chapter 9). In addition, there are a good many examples which extend the material of the text in a more or less informal way.

Material which is not directly relevant to the logical development is indicated by a bold asterisk ★ at the beginning and the end of the material,[†] or occasionally by some cautionary words at the outset. Thus the digressions in question can be regarded as optional reading, but even the reader intent on getting to the results may find the more relaxed atmosphere of some of the starred items congenial. Occasional use is made there of concepts and theorems which are related to the material of the text, but whose full exposition would take us too far afield. As an example, the proof of Fáry's theorem on straightening planar graphs (Theorem 1.34) assumes several plausible (but profound) results of point-set topology. The inclusion of this theorem makes possible several useful discussions later in the book; I have been careful to warn the reader that these discussions, resting as they do on a result whose complete proof involves concepts not explained here, must remain to some extent 'informal'. Investigations of this kind, and others in which an admission of informality is made, are in fact reminiscent of the style of argument usually adopted in more advanced mathematical work. The tacit assumption is present, that the author (and hopefully the reader) could, if he were so minded, fill in all the details and make the arguments formal, but that to do so in print would be, in this conservation-conscious age, a misuse of paper.

The fourth way in which my book differs from others of its kind is more technical in nature. I use collapsing as a major tool, rather than the Mayer–Vietoris sequence, which appears in a minor role in Chapter 7. My reason for

---

[†] thus asterisks do not direct the reader to footnotes, the symbol for which is (as you see) †.

using collapsing is that it leads straight to the calculation of the homology groups of surfaces with only a minimal need for invariance properties.

The theory of abelian groups enters into homology theory in two ways. The very definition of homology groups is framed in terms of quotient groups, and many theorems are expressed by the exactness of certain sequences. The abelian group theory used in this book is gathered together in an Appendix, in which proofs will be found of all the results used in the text.

About three-quarters of the material in this book was included in a course of around forty lectures I gave to third year undergraduates at Liverpool University. Included in the forty lectures were a few given by the students themselves. Some of these were on topics which now appear as optional sections, while others took their subject matter from research articles. The material now included in the text and which I omitted from the course comprises some of the work on planar graphs, all the work on invariance, the homology sequence of a triple, a few of the applications in Chapter 9, and several small optional sections. There are many other ways of cutting down the material. For example it is possible to restrict attention entirely to two-dimensional simplicial complexes: to facilitate this suggestion I have tried to ensure that by judicious skipping in Chapter 3 all mention of complexes of dimension higher than two can be suppressed. Thereafter the results can simply be restricted to this special case, and in Chapter 9 most of the work is two-dimensional anyway. Alternatively, Chapters 5 and 7 can be omitted, together with optional material according to the taste (or distaste) of the lecturer. I think it would be completely mistaken to give a course in which the students sharpened their teeth on all the techniques, only to leave without a sniff at the applications: better to omit details from Chapters 1–8 in order to include at any rate a selection of material from Chapter 9.

I have kept two principles before me, which I think should apply to all university courses in mathematics, and especially ones at a fairly high level. Firstly the course should have something to offer for those students who are going to take the subject further – it should be a stepping-stone to 'higher things'. But secondly, and I think this is sometimes forgotten, the course should exhibit a piece of mathematics which is reasonably complete in itself, and prove results which are interesting for their own sake. I venture to hope that the study of graphs in surfaces is interesting for its own sake, and that such a study provides a good introduction to the powerful and versatile techniques of algebraic topology. Bearing in mind that a proportion of students will become schoolteachers, I have been careful to include a number of topics, such as planar graphs in Chapter 1, colouring theorems in Chapter 2 and 'Brussels Sprouts' in Chapter 9, which, suitably diluted and imaginatively presented, could stimulate the interest of schoolchildren.

I have found it impossible in practice to separate worked examples, which often have small gaps in the working, from exercises, which often have hints for solution. All these are gathered together under the heading of 'Examples', but to help the reader in selecting exercises I have marked by X those examples where a substantial amount is left to the reader to verify. Two X's indicate an example that I found difficult. Some examples are starred: this indicates that their subject matter is outside the main concern of the book.

References to articles are given by the name of the author followed by the date of publication in brackets. A list of references appears at the end of the text. Some of the references within the text are to research articles and more advanced textbooks where the reader can find full details of subjects which cannot be explored here. Inevitably these references are often technical and not for the beginner unless he is very determined. I have also included among the references several books which, though they are not specifically referred to in the text, can be regarded as parallel reading.

There is a list of notation on page xvii. My only conscious departure from generally received notation is the use, in these days a little eccentric, of the declaration Q.E.D. to announce the end of a proof[†].

[†] For the present edition I have succumbed to the temptation to use the open square □ since this can flexibly indicate the end or the absence of a proof, and also the end of a statement whose proof preceded it.

Notation

I *Symbols* (Completely standard set-theoretic notation is not listed.)

Symbol	Example	Use	Page
juxtaposition	$s_p s_q$, KL	join	86
	K^r	r-skeleton of K	77
	fx	value of f at x	
	na	n integer, $a \in$ group	216
$\Rightarrow$		implies	
$\Leftrightarrow$		if and only if	
\	$X \setminus Y$	difference of sets	
$\searrow^e$	$K \searrow^e L$	elementary collapse	89
$\searrow$	$K \searrow L$	collapse	89
/	B/A	quotient group	220
|	$\delta_1 \mid \delta_2$	δ_1 is a factor of δ_2	231
( )	(vw)	edge	11
	$(v^0 \cdots v^p)$	p-simplex, p-chain	70, 100
	$K^{(r)}$	rth barycentric subdivision	135
(,)	(v, w)	edge	11
	(K, L)	pair	139
(; ,)	$(K; L_1, L_2)$	triad	159
(, ,)	(M, N, K)	triple	168
{ }	$\{z\}$, $\{z\}_K$, $\{z\}_{K,L}$	homology class	105, 114, 174
{ , }	$\{v, w\}$	edge	10
$\langle , \rangle$	$\langle \gamma, c \rangle$	value of cochain on chain	36
$\langle \ \rangle$	$\langle a \rangle$	subgroup generated by a	237

List of notation

Symbol	Example	Use	Page
$\approx$	$M \approx N$	equivalence of closed surfaces	53
$\cong$	$A \cong B$	isomorphism of groups	217
#	$M \# N$	connected sum	54
$\bullet$	$\overset{\bullet}{s}_n$	boundary of simplex	71, 76
$\circ$	$\overset{\circ}{s}_n$	interior of simplex	71
$<$	$s_p < s_n$	face of	72
$|\ |$	$|K|$	underlying space of	74
	$|A|$	number of elements of	216
$-$	$\bar{s}_n$	closure of simplex	76
	$\bar{\Sigma}, \overline{\mathrm{St}}$	closure of subset or star	84, 85
	$\bar{b}$	coset	220
	$\overline{M}$	closed surface obtained from surface	188
$\widehat{\ }$	$(v^0 \cdots \widehat{v^i} \cdots v^p)$	simplex with vertices $v^0 \cdots v^p$ except v^i	101
	$\widehat{\partial}$	relative boundary	113
	$\widehat{s}$	barycentre of	136
$\sim$	$z \sim z'$	homologous to	105
	$\tilde{H}_0$	reduced homology group	104
	$\tilde{i}_*$	reduced homomorphism	140
	$\tilde{A}$	vector space from abelian group	233
	$\tilde{f}$	linear map from homomorphism	234
$\sim$(mod 2)	$z \sim z'$(mod 2)	homologous mod 2 to	174
$\overset{L}{\sim}$	$z \overset{L}{\sim} z'$	homologous mod L	114
$\oplus$	$A \oplus B$	direct sum	222, 223
$\otimes$	$A \otimes B$	tensor product	179, 233
$'$	K'	barycentric subdivision	136
$''$	K''	second barycentric subdivision	183
$+$	V_i^+	region determined by V_i	183

II *Roman characters*

B	$B_p(K)$	boundary group	104, 124
	$B_p(K, L)$		114
	$B_p(K; 2)$		174
	$B_p(K, L; 2)$		176

List of notation

Symbol	Example	Use	Page
C	$C_p(K)$	chain group	21, 99, 124
C	$C_p(K,L)$		113
	$C_p(K;2)$		171
	$C_p(K,L;2)$		176
	$C^p(G)$	cochain group	36
	CK	cone on K	145
Cl	$\text{Cl}(\Sigma, K)$	closure	84
F	FS	free abelian group on S	219
$\mathscr{F}$	$\mathscr{F}(R)$	frontier of region	30
$GL(n, \mathbb{R})$		general linear group	98
H	$H_p(K)$	homology group	104, 124
	$H_p(K,L)$		114
	$H_p(K;2)$		174
	$H_p(K,L;2)$		176
	$\tilde{H}_0(K), \tilde{H}_0(K;2)$	reduced homology group	104, 174
$\mathcal{H}$	$\mathcal{H}(x)$	Heawood number	64
i_*		homomorphism from inclusion	121
$\tilde{i}_*$		reduced homomorphism	140
Im	$\text{Im} f$	image	104, 217
j, j_q		homomorphism forgetting a subcomplex	113
j_*		homomorphism from j	122
$\tilde{j}_*$		reduced homomorphism	140
Ker	$\text{Ker} f$	kernel	104, 217
Lk	$\text{Lk}(s, K)$	link	87
N		regular neighbourhood	183
P, kP		projective plane, k-fold projective plane	54
$\mathbb{Q}$		rational numbers	218
$\mathbb{R}^n$		real euclidean or vector space of dimension n	
S		sphere	54
S^n		n-sphere	148
St, $\overline{\text{St}}$	$\text{St}(s, K)$	star, closed star	85
T, hT		torus, n-fold torus	54
Tor	$\text{Tor}(A, B)$	torsion product	179

xx *List of notation*

Symbol	Example	Use	Page
V		subcomplex complementary to regular neighbourhood	183
w_p		Stiefel–Whitney class	178
Z	$Z_p(K)$	cycle group	104, 124
	$Z_p(K,L)$		114
	$Z_p(K;2)$		174
	$Z_p(K,L;2)$		176
$\mathbb{Z}$		additive group of integers	216
$\mathbb{Z}_k$		$\{0, 1, \ldots, k-1\}$ under addition modulo k	216
$\mathbb{Z}^n$		group of n-tuples of integers	218

III. *Greek characters*

α, α_p		homomorphism	129, 167
	$\alpha_p(K), \alpha_p(K,L)$	number of p-simplexes of $K, K\backslash L$	55, 91, 155
β, β_p		homomorphism	130, 168
	$\beta_p(K)$	Betti number	155
	$\beta_p(K,L)$		155
	$\widehat{\beta}_p(K)$	connectivity number	174
γ, γ_p		homomorphism	168
δ		coboundary	36
∂, ∂_p		boundary homomorphism	21, 101, 172
∂	∂M	boundary of surface M	187
$\widehat{\partial}, \widehat{\partial}_p$		relative boundary	113
∂_*		homomorphism from ∂	123
$\widetilde{\partial}_*$		reduced homomorphism from ∂	140
Δ		homomorphism	159
	$\mathscr{G}^\Delta$	triangular graph	32
ε		augmentation	25, 103, 172
μ		cyclomatic number	19
ρ		restriction homomorphism	116
ϕ		homomorphism	158
χ	$\chi(K)$	Euler characteristic	55, 155
	$\chi(K,L)$		155
ψ		homomorphism	159

Introduction

Algebraic methods have often been applied to the study of geometry. The algebraic reformulation of euclidean geometry as 'coordinate geometry' by Descartes was peculiarly successful in that algebra and geometry exactly mirrored each other. The objects studied in this book – graphs, surfaces and higher dimensional 'complexes' – are much more complicated than the lines and circles of euclidean geometry and it is not to be expected that any reasonable algebraic machinery can hope to describe all their features. We must expect many distinct geometrical situations to be described by the same piece of algebra. That is not necessarily a disadvantage, for not all features are equally interesting and it can be useful to have a systematic way of ignoring some of the less interesting ones.

Throughout the book we shall be concerned with subsets X of a euclidean space (such as ordinary space $\mathbb{R}^3$) and shall attempt to clarify the structure of the sets X by looking at what are called *cycles* on them. For the present let us consider one-dimensional cycles, or 1-cycles, which have their origin in the idea of simple closed curve (that is, a closed curve without self-intersections). Thus consider the three graphs drawn below. It is fairly evident that no simple closed curves at all can be drawn on the left-hand graph.

On the middle graph a simple closed curve can be drawn going once round the triangle, while on the right-hand graph there are three triangles, two right-angled and one equilateral, which can be circumnavigated. The first hint of algebra comes when we try to combine two cycles (in this case two simple closed curves) together. We want to say something like

The way in which we make the vertical edge cancel out is by giving orientations (directions) to the edges of the triangles and taking all cycles to be, say, clockwise. Then we have

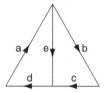

$$a + b + c + d = (a + e + d) + (b + c - e)$$

where the $-e$ occurs because e is going the wrong way for the right-hand cycle. The sum now looks like a calculation in an abelian group, and indeed in Chapter 1 we make the set of 1-cycles into an abelian group which measures the 'number of holes' in the graph, or the 'number of independent loops' on the graph. This number is respectively 0, 1 and 2 for the three graphs above, and the corresponding groups are 0, $\mathbb{Z}$, $\mathbb{Z} \oplus \mathbb{Z}$. The amount of group theory needed to read Chapter 1 is minimal, barely more than the definitions of abelian group and homomorphism.

When we come to consider 1-cycles on higher-dimensional objects the situation becomes more interesting. The diagrams

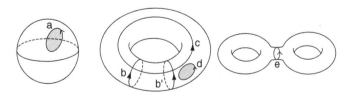

represent a spherical surface, a torus (or 'inner tube') and a double torus or pretzel, and a, b, b', c, d, e are 1-cycles. Now a and d are each the boundary

Introduction

of a region – such a region is shaded, but it would also be possible to use the 'complementary' region occupying the remainder of the surface. On the other hand b, b' and c do not bound regions. The fact that the simple closed curve b fails to bound a region reveals the presence of a hole in the torus (where the air goes in an inner tube): 'b goes round the hole'. Similarly c reveals the presence of another hole (where the axle of the wheel goes). Since b and c are one-dimensional we could call the holes one-dimensional too.

In order to use cycles for gathering geometric information we shall declare those which, like a and d, bound a region, to be 'zero' – homologous to zero is the official phrase, denoted ~ 0. Now b and b' between them bound a region; once the idea of orientation is made precise this implies that $b - b' \sim 0$ (rather than $b + b' \sim 0$). Rearranging, we have $b \sim b'$, which gives formal expression to the fact that b and b' go round the same hole in the torus. There is a technique available for making the bounding cycles zero: they form a subgroup (the *boundary* subgroup) of the group of cycles and what is required is the quotient group {cycles}/{boundaries}. This quotient group, when the cycles are 1-cycles, is called the *first homology group*. The complexity ('number of free generators') of this group measures, roughly speaking, the number of one-dimensional holes the object has. The sphere has none and the torus has two; their first homology groups are respectively 0 and $\mathbb{Z} \oplus \mathbb{Z}$.

Notice in passing that a and d can each be shrunk to a point on its respective surface whereas e, on the double torus, cannot. Nevertheless e does bound a region (namely one half of the surface) so $e \sim 0$.

The whole of, for example, the spherical surface can be considered as a two-dimensional cycle, or 2-cycle, since it is 'closed'; certainly this cycle is not the boundary of anything since there is nothing three-dimensional *in the surface* for it to bound. If on the other hand we replace the sphere by the solid ball obtained by filling in the inside of the sphere, then the 2-cycle given by the sphere becomes a boundary: in the solid ball this 2-cycle is a boundary. As before we can form the homology group, this time the *second homology group*; for the sphere this is $\mathbb{Z}$ and for the solid ball it is 0.

Notice that if we consider a solid torus (like a solid tyre) then b and b' become homologous to zero, while c does not. In fact the first homology group changes from $\mathbb{Z} \oplus \mathbb{Z}$ to $\mathbb{Z}$.

It is important to realize that the homology groups of an object depend only on the object and not on the way in which it is placed in space. Thus if we cut the torus along b, tie it in a knot and then re-stick along b, the resulting 'knotted torus' has the same homology groups as the original one. Stepping down one dimension we could consider the same graph being placed in different ways in a surface, such as the torus in the diagrams below:

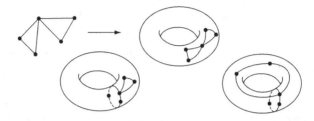

These three situations are quite different – for instance the number of regions into which the torus is divided is different in each case. We can study 'embedding' or 'placement' problems by means of *relative homology groups*, which take into account both graph and torus at the same time. The ordinary groups and the relative groups are introduced in Chapter 4.

Here are some of the problems which are studied in Chapter 9 by means of relative homology groups.

1. Given a graph in a surface, what can be said about the number of regions into which the graph divides the surface and about the nature of those regions? Much of Chapter 9 is concerned with this problem.

2. What is the 'largest' graph which can be embedded in a surface without separating it into two regions? Of course it is necessary to choose a good definition of largeness first; here is an alternative formulation of the question. How many simple closed cuts can be made on a surface before it falls into two pieces? For example on a double torus the answer is four – see the diagram on the left. For a 'Klein bottle', obtained from a plane rectangle by identifying

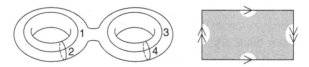

the sides in pairs as indicated on the diagram on the right (compare 2.2(4)), the answer is two. Can the reader find two suitable curves? The general answer is in 9.20. The Klein bottle cannot be constructed in three-dimensional space, and a proof that this is so is sketched in 9.18. It can be constructed in $\mathbb{R}^4$ (2.16(5)).

3. What can be said about a cycle given by a simple closed curve on, say, the torus? The simple closed curve labelled b in the middle diagram on page 2 winds round one way and the simple closed curve c winds round the other way. Can a simple closed curve wind round, say, twice the b-way and not go round

the c-way at all? Certainly the most obvious candidates are not *simple* closed curves:

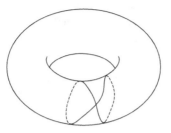

The general answer is in 9.22.

Curiously, although cycles were studied from the beginnings of algebraic topology in the writings of H. Poincaré at the end of the nineteenth century, no explicit mention was made of homology *groups* until much later, around the early 1930s. What were used instead were numerical invariants of the groups (the so-called rank and torsion coefficients; see A.26), which accounts for the lack of formal recognition given to many of the basic results. For example the exact homology sequence of a pair (Chapter 6) did not appear explicitly until 1941. The reformulation of homology theory in terms of abelian groups not only produced a substantial increase in clarity but made available powerful tools with which the scope of the theory could be broadened.[†]

This is an example – one of many – where modern algebraic formalism has come to the rescue of an attractive but elusive geometrical idea. Of course any rescue operation has its dangers; in this case that formalism should cease to be subservient to geometry and become an end in itself. 1 have tried very hard to avoid such a course and to keep constantly in mind that the goal of the book is the geometrical applications of Chapter 9.

The techniques developed in this book do not enable us to study 'curved surfaces' such as the torus or sphere directly. (The techniques of differential topology are more suited to such a direct study; for a beautiful introduction to that subject see the book of Milnor listed in the References.) Instead we 'triangulate' these surfaces, that is, divide them up into (for the moment curved) triangular regions in such a way that two regions which intersect do so either at a single vertex or along a single edge. Triangulating a surface reduces the calculation of homology groups to a finite procedure, which is explained in Chapter 4; it has the additional advantage that it is possible to give a complete prescription for constructing triangulations of surfaces. This is given in Chapter 2; as

---

[†] For more historical information see the books by Dieudonné, Lakatos *et al.*, Kline, Eilenberg & Steenrod and Biggs, Lloyd & Wilson, and the article by Hilton (1988) listed in the References.

an example consider the diagrams below, showing a tetrahedron and an octahedron each inside a sphere. The hollow tetrahedron can be projected outwards ($P \to P'$) on to the sphere which surrounds it, thus giving a triangulation of the sphere. Likewise the hollow octahedron can be projected outwards ($P \to P'$) to give another triangulation. Indeed, the hollow tetrahedron and octa-

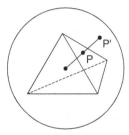 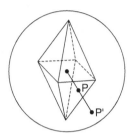

hedron can themselves be regarded as triangulated surfaces in which it so happens that the triangles are flat and the edges straight. These rectilinear triangulations are referred to as *simplicial complexes*; they are studied in considerable generality in Chapter 3. Their importance lies in the fact that any triangulated surface is homeomorphic to one in which the triangles are flat and the edges straight. (Two subsets X and Y of euclidean space are called *homeomorphic* if there exists a continuous, bijective map $f : X \to Y$ with continuous inverse f^{-1}. Such a map f is called a *homeomorphism*.) The maps $P \to P'$ above are both homeomorphisms. The kind of geometrical information which we are trying to capture – that is, information of a topological nature – is so basic that it is unaltered by homeomorphisms. Thus it is no restriction to concentrate entirely on rectilinear triangulations.

That is what we do in this book: all triangulations have flat triangles with straight edges. It should be noted, however, that often when drawing surfaces we shall draw them curved. This is for two reasons: the pictures are more recognizable and easier to draw; and they help to remind us that the triangulations are a tool to help us gather information, not an intrinsic part of the geometrical data.

The remarkable thing is that homology groups, in spite of being defined via triangulations, really do measure something intrinsic and geometrical. To be more precise: if X and Y are homeomorphic, then the homology groups of X are isomorphic to those of Y. This is true not only if X and Y are surfaces, but if they are any simplicial complexes which may include, besides triangles, solid tetrahedra and higher dimensional 'simplexes'. This result is called the *topological invariance of homology groups*, and it has the following equivalent formulation: if, for some p, the pth homology group of X is not isomorphic to the pth homology group of Y, then X and Y are not homeomorphic. Thus homology groups can be used to distinguish objects from each other. It is not true in general that if the homology groups of X and Y are isomorphic then X and Y are homeomorphic: many different objects can have the same groups. This brings us back to the point made at the beginning of this Introduction, that many distinct geometrical situations are described by the same piece of algebra, and shows that homology groups are not the ultimate tool in distinguishing objects topologically from one another.

It happens that the applications in this book do not require the full invariance theorem quoted above, and accordingly this theorem is not proved here. Enough, in fact more than enough, is proved in Chapter 5 for our purposes. A sketch of the argument leading to the theorem of topological invariance is also included, together with references for the full proof.

The finite procedure for calculating homology groups can in practice be very lengthy, and for this reason general theorems are a help. A number of such theorems are given in Chapter 6; these enable the homology groups of all (closed) surfaces to be calculated without difficulty. Two more theorems, slightly more technical in nature and not strictly necessary for the applications, are given in Chapter 7.

Some of the work of Chapter 9 needs an 'unoriented' homology theory which differs in its details from the oriented homology theory studied in Chapters 1–7. This is presented in Chapter 8.

An Appendix contains all the group theory needed to read the book, presented in a fairly condensed form. Virtually no group theory is needed for Chapters 1–3.

Finally it should be emphasized that there are other approaches to homology theory. It is possible to define homology groups for far more general objects than simplicial complexes, in fact for arbitrary 'topological spaces' which may not be contained in a euclidean space at all. There are several methods for doing this, described as 'singular' or 'Čech' or 'Alexander–Spanier' homology. There is also a relatively recent development called 'Intersection Homology', designed to apply to spaces with singularities. These all have the property of topological invariance, so that they can sometimes be used to distinguish two spaces from one another – indeed this is one of the principal applications of homology theory. Again the method is not infallible: for example the three objects pictured below (cylinder, Möbius band, circle) all have the same groups in any homology theory, but no two are homeomorphic. Accounts of other theories can be found in, for example, the books of Spanier and of Eilenberg & Steenrod listed in the References. The latter classic text also presents a most attractive account of homology theory from the axiomatic point of view.

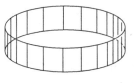

Cylinder

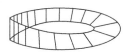

Möbius band

Circle

1
Graphs

The theory of graphs, otherwise known as networks, is a branch of mathematics which finds much application both within mathematics and in science. There are many books where the reader can find evidence to support this claim, for example those by Berge, Gross & Yellen, Diestel, Wilson & Beinecke and Bollobás listed in the References. Our motivation for touching on the subject here is different. Many of the ideas which we shall encounter later can be met, in a diluted form, in the simpler situation of graph theory, and that is the reason why we study graphs first. The problems studied by graph theorists are usually specifically applicable to graphs, and do not have sensible analogues in the more general theory of 'simplicial complexes' which we shall study. It is hardly surprising, therefore, that the theory we shall develop has little to say of interest to graph theorists, and it is partly for this reason that an occasional excursion is made, in this chapter and elsewhere, into genuine graph-theoretic territory. This may give the reader some idea of the difficult and interesting problems which lie there, but which the main concern of this book, namely homology theory, can scarcely touch.

Abstract graphs and realizations

A graph is intuitively a finite set of points in space, called the vertices of the graph, some pairs of vertices being joined by arcs, called the edges of the graph. Two arcs are assumed to meet, if at all, in a vertex, and it is also assumed that no edge joins a vertex to itself and that two vertices are never connected by more than one edge. (Such a graph is often called a *s*imple graph.) That is, we do not allow

The positions of the vertices and the lengths of the edges do not concern us; what is important is the number of vertices and the pairs of vertices which are connected by an edge. Thus the only information we need about an edge is the names of the vertices which it connects. This much information can be summed up in the following 'abstract' definition; we shall return to 'reality' shortly.

Definition 1.1. An *abstract graph* is a pair (V, E) where V is a finite set and E is a set of unordered pairs of distinct elements of V. Thus an element of E is of the form $\{v, w\}$ where v and w belong to V and $v \neq w$. The elements of V are called *vertices* and the element $\{v, w\}$ of E is called the *edge* joining v and w (or w and v).

Note that E may be empty: it is possible for no pair of vertices to be joined by an edge. Indeed V may be empty, in which case E certainly is – this is the empty graph with no vertices and no edges. In any case if V has n elements then E has at most $\binom{n}{2} = \frac{1}{2}n(n-1)$ elements. An abstract graph with the maximum number of edges is called *complete*: every pair of vertices is joined by an edge.

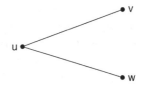

As an example of an abstract graph let $V = \{u, v, w\}$ and $E = \{\{u, v\}, \{u, w\}\}$. This abstract graph (V, E) can be pictured, or 'realized' as we shall say in a moment, by the diagram.

In many circumstances it is convenient to speak of graphs in which each edge has a direction or orientation (intuitively an arrow) attached to it. For example in an electrical circuit carrying direct current each edge (wire) has a direction, namely the direction in which current flows.

Definition 1.2. An *abstract oriented graph* is a pair (V, E) where V is a finite set and E is a set of ordered pairs of distinct elements of E with the property that, if $(v, w) \in E$ then $(w, v) \notin E$. The elements of V are called *vertices* and the element (v, w) of E is called the *edge* from v to w.

Associated with any abstract oriented graph there is an abstract graph, obtained by changing ordered to unordered pairs. The two graphs have the same number of vertices, and the same number of edges on account of the assumption that (v, w) and (w, v) never both belong to E.

We now return to graphs 'realized' in space.

Definition 1.3. Let (V, E) be an abstract graph. A *realization* of (V, E) is a set of points in a real vector space $\mathbb{R}^N$, one point for each vertex, together with straight segments joining precisely those pairs of points which correspond to edges. The points are called *vertices* and the segments *edges*; the realization is called a *graph*. We require two 'intersection conditions' to hold: (i) two edges meet, if at all, in a common end-point, and (ii) no vertex lies on an edge except at one of its end-points. The edge joining v and w will be denoted (vw), or (wv).

Note that the edges of a graph are straight; that this is no restriction for graphs in $\mathbb{R}^3$ follows from Theorem 1.4 below. We shall turn to the question of whether it is a restriction for graphs in $\mathbb{R}^2$ (planar graphs) in 1.34.

A realization of an abstract oriented graph, called an *oriented graph*, is just a realization of the associated abstract graph, together with the ordering on each edge of the graph. In practice this ordering is indicated by an arrow from the first to the second vertex of the edge:

is the edge from v to w, which will be denoted (vw).

The complete abstract graphs with up to four vertices have realizations in $\mathbb{R}^2$:

The reader can probably convince himself that the complete abstract graph with five vertices cannot be realized in $\mathbb{R}^2$ (see Proposition 1.32); however it can be realized in $\mathbb{R}^3$, as indeed can every abstract graph.

Theorem 1.4. *Every abstract graph can be realized in $\mathbb{R}^3$.*

Proof Let C be the subset of $\mathbb{R}^3$ defined by

$$C = \{(x, x^2, x^3) : x \in \mathbb{R}\}.$$

(C is called a 'twisted cubic'.) I claim that no four points of C are coplanar (it follows from this that no three points of C are collinear). To prove the claim take four points P_1, P_2, P_3, P_4 of C, given respectively by distinct real numbers x_1, x_2, x_3, x_4. They are coplanar if and only if the vectors $P_1 - P_2$, $P_1 - P_3$, $P_1 - P_4$ are linearly dependent,

i.e. $\begin{vmatrix} x_1 - x_2 & x_1^2 - x_2^2 & x_1^3 - x_2^3 \\ x_1 - x_3 & x_1^2 - x_3^2 & x_1^3 - x_3^3 \\ x_1 - x_4 & x_1^2 - x_4^2 & x_1^3 - x_4^3 \end{vmatrix} = 0.$

However this determinant vanishes if any only if the x's are not distinct, and this proves the claim.

To realize an abstract graph (V, E) in $\mathbb{R}^3$ take n distinct points $P_1, \ldots, P_n$ of C, one for each vertex, and join precisely those pairs of points which correspond to elements of E by straight segments. It remains to verify that the intersection conditions hold. Let $P_i P_j$ and $P_k P_\ell$ be two (distinct) segments; certainly $i \neq j$ and $k \neq \ell$ and we can assume $j \neq \ell$. If $i = k$, so that the segments have a common end-point, it is still true that P_i, P_j and P_ℓ are distinct and so not collinear; hence the segments do not have any more points in common. If on the other hand $i \neq k$ then all four P's are distinct and so not coplanar; hence the segments do not meet at all. This shows that (i) of 1.3 holds; (ii) follows immediately from the fact that no three points of C are collinear. □

Remarks 1.5. (1) When drawing graphs it is often convenient to draw a figure in the plane with heavy dots for vertices and straight lines (or even curved ones) for edges, possibly intersecting at points other than end-points. For example the complete graph with five vertices might be drawn as in the diagram below. This is to be regarded as a recipe for constructing an abstract graph which Theorem 1.4 assures us can be realized in $\mathbb{R}^3$. When the edges are straight, as in the diagram, it can also be regarded as the projection into the plane of a realization in $\mathbb{R}^3$.

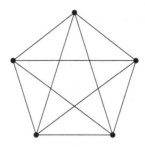

(2) The proof of 1.4 still works if the definition of 'abstract graph' is relaxed to the extent of allowing the set V of vertices to be countable.

Of course the same abstract graph can be realized in many ways. The graph in $\mathbb{R}^3$ formed by the edges and vertices of a cube, and the planar graph (see the diagram below) are realizations of the same abstract graph. This, and the fact that the precise names given to the vertices of an abstract graph are unimportant, suggests the following definition.

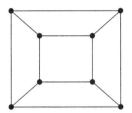

Definition 1.6. Two abstract graphs (V, E) and (V', E') are called *isomorphic* if there is a bijective map $f : V \to V'$ such that

$$\{v, w\} \in E \Leftrightarrow \{f(v), f(w)\} \in E'.$$

(Thus f is simply a relabelling of vertices which preserves edges.) Two graphs are called *isomorphic* if they are realizations of isomorphic abstract graphs.

The corresponding definitions for oriented graphs are obtained by replacing unordered pairs { , } by ordered pairs (,).

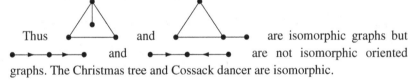

Thus ... and ... are isomorphic graphs but ... and ... are not isomorphic oriented graphs. The Christmas tree and Cossack dancer are isomorphic.

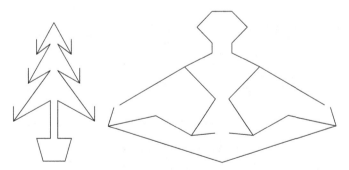

(For once, vertices are not emphasized with dots.)

To prove two graphs are isomorphic we have to construct an explicit map from the vertex set of one to the vertex set of the other which is bijective and preserves edges. To prove two graphs are not isomorphic we look at properties of graphs which are invariant under isomorphisms and try to find such a property possessed by one graph but not by the other. The simplest such property is the number of vertices or edges; next simplest is the orders of the vertices. The *order* of a vertex is the number of edges with the vertex as an end-point, and clearly two isomorphic graphs must have the same number of vertices of any given order. Slightly more subtle methods are needed for the following two examples.

Examples 1.7. (1X) Exactly two of the following three graphs are isomorphic. Which two? (Compare remark 1.5(1).)

[Hint. Look at the vertices of order 3.]

(2X) The first two graphs below are isomorphic, and the vertices are labelled to show this. Which other pairs are isomorphic? (Again see 1.5(1).)

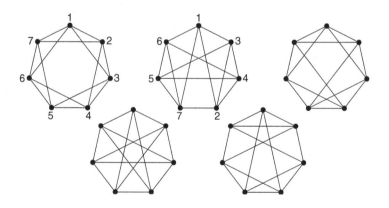

[Hint. Count triangular paths.]

★ Kirchhoff's laws

The discussion which follows is intended to show how one of the fundamental properties of a graph, its 'cyclomatic number', arises naturally in the study of direct current (D.C.) electrical circuits. The development of graph theory proper continues at 1.8.

Kirchhoff's laws

We can associate a graph with a D.C. circuit, replacing terminals by vertices and wires by edges. An edge joining a vertex to itself or two or more edges connecting the same pair of vertices can easily be eliminated by the addition of extra vertices which do not essentially change the graph from an electrical point of view. We can also assume that the graph is all in one piece, since separate D.C. circuits can be treated separately. Each edge has associated with it positive real number called the resistance, and some edges will in addition contain sources of power. The edges are oriented (arbitrarily) and the object is to calculate the current in each edge; a negative answer indicates that the current is flowing against the chosen orientation.

Kirchhoff's laws provide linear equations for the currents. These equations fall into two types:

(K1) There is one equation for each closed loop† in the graph. A loop is first given a direction, one way round or the other, and in the resulting equation the only currents which appear are those along edges in the loop. The coefficient of a current is the resistance of the edge if the orientation of the edge agrees with that of the loop; otherwise it is minus the resistance. The 'right-hand side' of the equation is the total electro-motive force (e.m.f.) in the loop, with an appropriate convention on signs.

(K2) There is one equation for each vertex, expressing the fact that the total current flowing into the vertex is zero (a current flowing out being counted negatively).

Suppose that there are α_0 vertices and α_1 edges; then the α_0 equations (K2) are not independent since if we add them all up the sum vanishes identically. However it is not hard to verify that any $\alpha_0 - 1$ of them are independent. Since there are α_1 unknowns it follows that in order to determine the currents we shall need at least $\alpha_1 - \alpha_0 + 1$ independent equations (K1) – possibly more, since there may be relations between these and the (K2) equations. In fact, there are exactly $\alpha_1 - \alpha_0 + 1$ independent equations (K1) (that is, $\alpha_1 - \alpha_0 + 1$ 'independent loops') and these are also independent of (K2) and so (K1) and (K2) between them uniquely determine the currents. A proof of these facts is sketched in an appendix at the end of the chapter, where we have more notation at our disposal. In the meantime here is a simple proof for the case of a graph with a single loop. For convenience assume the edges $e^1, \ldots, e^n$ of the loop are all oriented in the same direction round the loop. Let $v^1, \ldots, v^n$ be the vertices round the loop; there may be other vertices and edges not in the loop but these do not affect the argument.

---

† Here we use 'loop' to mean a simple closed path round the edges; a precise definition is in 1.8 below. Beware that some authors use 'loop' for an edge joining a vertex to itself, something we forbid here.

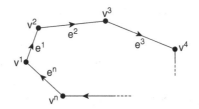

Suppose that the (K1) equation of the loop is a linear combination of (K2) equations, the coefficient of the (K2) equation for v^i being $\lambda_i, i = 1, \ldots, n$. The current along edge e^1 can occur only in the (K2) equations for v^1 and v^2, so that the coefficient of this current in the (K1) equation of the loop must be $\lambda_2 - \lambda_1$. Since this is positive (resistances are positive and e^1 is oriented round the loop) we have $\lambda_2 > \lambda_1$. In a similar way $\lambda_3 > \lambda_2, \lambda_4 > \lambda_3, \ldots, \lambda_n > \lambda_{n-1}, \lambda_1 > \lambda_n$. Hence $\lambda_1 > \lambda_1$, a contradiction.

The 'number of independent loops' $\alpha_1 - \alpha_0 + 1$ arising from Kirchhoff's analysis of electrical circuits was called the 'cyclomatic number' of the circuit by James Clerk Maxwell.

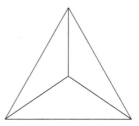

As a simple example, the graph above has cyclomatic number 3 and it is certainly very plausible that every loop is a 'combination' of the three loops which bound the regions inside the big triangle. ★

Maximal trees and the cyclomatic number

Definitions 1.8. A *path* on a graph G from v^1 to v^{n+1} is a sequence of vertices and edges

$$v^1 e^1 v^2 e^2 \cdots v^n e^n v^{n+1}$$

where $e^1 = (v^1 v^2), e^2 = v^2 v^3, \ldots, e^n = (v^n v^{n+1})$. (See 1.3 for notation. The path goes from v^1 to v^{n+1} along the edges $e^1, \ldots, e^n$.) Here $n \geq 0$: for $n = 0$ we just have v^1 and no edges. If G is oriented, we only require that $e^i = (v^i v^{i+1})$ or $(v^{i+1} v^i)$ for $i = 1, \ldots, n$; that is, the edges along the path do not have to point in the 'right' direction. The path is called *simple* if the edges $e^1, \ldots, e^n$ are all distinct, and the vertices $v^1, \ldots, v^{n+1}$ are all distinct except that possibly $v^1 = v^{n+1}$. If the simple path has $v^1 = v^{n+1}$ and $n > 0$ it is called a *loop*; thus in fact a loop always has $n \geq 3$. We shall regard two loops as equal when they consist of the same vertices and the same edges.

A graph G is called *connected* if, given any two vertices v and w of G there is a path on G from v to w. For any non-empty graph G, a *component* of G consists of all the edges and vertices which occur in paths starting at some particular vertex of G. Thus a non-empty connected graph has exactly one component. A graph which is connected and has no loops is called a *tree*.

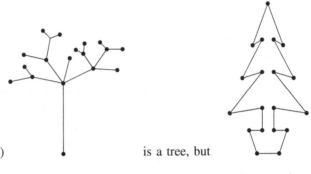

Examples 1.9. (1) is a tree, but is not.

(2) Given n points in space – say n points on the twisted cubic of the proof of 1.4 – we can ask how many trees there are with those points as vertices. For instance when $n = 3$, the answer is 3:

However these are all isomorphic, and a more interesting question is: how many trees are there no two of which are isomorphic? The former question has the answer n^{n-2} ($n \geq 1$), but the general answer to the latter question is unknown. (The first result is known as Cayley's Theorem, and it is discussed in books on graph theory such as those listed at the beginning of this chapter.)

(3) If there is a path on G from v to w then by elimination of any repetitions a simple path can be constructed on G from v to w.

(4X) An edge e of a graph is part of some loop on the graph if and only if the removal of e does not increase the number of components of the graph.

Definitions 1.10. Given a graph G, a graph H is called a *subgraph* of G if the vertices of H are vertices of G and the edges of H are edges of G. Also H is called a *proper subgraph* of G if in addition $H \ne G$. Any graph G will have a subgraph which is a tree (e.g. the empty subgraph is a tree!) so that the set $\mathscr{T}$ of subgraphs of G which are trees will have maximal elements. That is, there will be at least one $T \in \mathscr{T}$ such that T is not a proper subgraph of any $T' \in \mathscr{T}$. Such a T is called a *maximal tree* for G. (Some authors say *spanning tree*.)

Maximal trees are mainly of interest for connected graphs, and we have the following result.

Proposition 1.11. *Let G be a connected graph. A subgraph T of G is a maximal tree for G if and only if T is a tree containing all the vertices of G.*

Proof We may suppose G is non-empty. Suppose that T is a maximal tree for G. Certainly T is then a non-empty tree; suppose that a vertex v of G fails to belong to T. Choose a vertex w of T and a path from v to w on G. Let v^{i+1} be the first vertex of this path which belongs to T; then the previous vertex v^i and the edge $e^i = (v^i v^{i+1})$ will not belong to T. Adding v^i and e^i to T we obtain say T', a larger subgraph of G than T. Furthermore, e^i cannot be part of any loop on T' since one of its end-points, v^i, does not belong to any edge of T' other than e^i. Thus T' is a tree and T is not maximal.

Suppose conversely that T is a subgraph of G which is a tree and contains all the vertices of G. Suppose that H is a subgraph of G with T a proper subgraph of H. Since T already has all the vertices of G there must be an edge, $e = (vw)$ say, in H but not in T. Since T is connected there is a simple path on T from v to w (compare 1.9(3)); adding e to the end of this path gives a loop in H. Hence H is not a tree and T is maximal. □

Examples 1.12. (IX) Given a connected graph G and a subgraph T which is a tree, T is a subgraph of a maximal tree for G. Indeed, given a subgraph H of G which contains no loops (but is not necessarily connected), H is a subgraph of a maximal tree for G. (Use the method of the first paragraph of the proof of 1.11.)

(2X) A vertex v of an edge (vw) in a graph G is called *free* (or *terminal*) if v is not a vertex of any edge of G other than e. An *inner edge* of a graph is an edge neither of whose end-points is a free vertex.

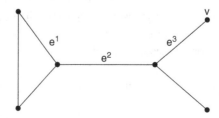

Thus in the diagram e^1 and e^2 are inner edges, e^3 is not and v is a free vertex of e^3. A graph with at least one edge, all of whose edges are inner, is bound to have a loop (since the number of edges is finite); it follows that a tree with at least one edge always possesses a non-inner edge. Removing a non-inner edge e of a graph, together with a free vertex of e, is called an *elementary collapse*, and a sequence of these is called a *collapse*. (We shall meet this idea again in 3.30.) Since a collapse takes a tree to a smaller tree it follows that any tree collapses to a single vertex. Indeed, the converse is true: if a graph collapses to a single vertex, then it must have been a tree.

The following definition and theorem connect the idea of maximal tree with the number $\alpha_1 - \alpha_0 + 1$ introduced during the discussion of Kirchhoff's laws.

Definition 1.13. Let G be a connected graph with $\alpha_0 = \alpha_0(G)$ vertices and $\alpha_1 = \alpha_1(G)$ edges. The *cyclomatic number* of G is the integer

$$\mu = \mu(G) = \alpha_1 - \alpha_0 + 1.$$

Theorem 1.14. *Let G be a graph with cyclomatic number μ. Then*

(1) $\mu \geq 0$.
(2) $\mu = 0$ *if and only if G is a tree.*
(3) *Every maximal tree for G has $\alpha_1 - \mu$ edges. That is, every maximal tree for G can be obtained from G by removing μ suitable edges (and no vertices). These edges must be inner edges since T is connected.*
(4) *Removing fewer than μ edges from G cannot produce a tree; removing more than μ edges always disconnects G.*

Proof The proof of (1) and (2) is by induction on α_1. Clearly (1) and (2) are true when $\alpha_1 = 0$; suppose they are true when $\alpha_1 < k$, where $k \geq 1$, and let G have k edges.

(i) Suppose G is a tree. Then G has a non-inner edge (compare 1.12(2)), e say. Removing e and a free vertex of e from G leaves a tree G' with $k - 1$ edges. By the induction hypothesis, $\mu(G') = 0$, and by construction $\mu(G') = \mu(G)$. Hence $\mu(G) = 0$.

(ii) Suppose G is not a tree. Then G has an inner edge, namely any edge of a loop of G; let G' be obtained by removing this edge (and no vertices). Removing an edge from a loop cannot disconnect G, so that G' is connected and by the induction hypothesis $\mu(G') \geq 0$. By construction $\mu(G') = \mu(G) - 1$; hence $\mu(G) \geq 1$.

This completes the proof by induction of (1) and (2).

To prove (3), let T be a maximal tree for G, so that by 1.11 $\alpha_0(G) = \alpha_0(T)$. Using this and $\mu(T) = 0$ it follows easily that $\alpha_1(T) = \alpha_1(G) - \mu(G)$.

To prove (4), note that removing p edges (and no vertices) from G leaves either a disconnected graph or else a connected graph with cyclomatic number $\mu(G) - p$. From this and (1) and (2) we can deduce (4). □

Remarks 1.15. (1) Removing μ edges at random may not leave a connected graph, hence certainly not a tree. However if it does leave a connected graph then this must be a maximal tree: it will contain all the vertices and have cyclomatic number $\mu - \mu = 0$.

(2) Let G be a connected graph with $\mu(G) = 1$. Then G has exactly one loop. (Proof: G has at least one loop since $\mu(G) > 0$. If there are two loops let e be an edge of one loop not in the other. Removing e from G leaves G connected, and still containing a loop, but it also lowers μ by one to zero. This is a contradiction.)

1.16. The basic loops

Let G be a connected graph with cyclomatic number μ and let $e^1, \ldots, e^\mu$ be edges whose removal from G leaves a maximal tree T (see 1.14 (3)). We shall denote by $T + e^i$ the graph obtained from T by replacing e^i. Then $\mu(T+e^i) = 1$, and by 1.15(2) there is a unique loop ℓ^i on $T + e^i$. We shall prove that in some sense these loops 'generate' all loops on G; one consequence of this is that the Kirchhoff equations for $\ell^1, \ldots, \ell^\mu$ are enough to determine all equations of type (K1). Before doing this we shall generalize the idea of loop.

Chains and cycles on an oriented graph

Let G be an oriented graph. Given a path

$$v^1 e^1 v^2 e^2 \ldots v^n e^n v^{n+1}$$

on G we can associate with it a formal sum

$$\varepsilon_1 e^1 + \varepsilon_2 e^2 + \cdots + \varepsilon_n e^n$$

where $\varepsilon_i = +1$ if $e^i = (v^i v^{i+1})$ and $\varepsilon_i = -1$ if $e^i = (v^{i+1} v^i)$. Thus the formal sum is obtained by travelling along the path from v^1 to v^{n+1} and writing down the edges encountered, with a minus sign if they are oriented against the direction of travel. More generally, we have the following.

Definition 1.17. Let G be an oriented graph with edges $e^1, \ldots, e^r$. A *1-chain* on G is a formal sum

$$\lambda_1 e^1 + \lambda_2 e^2 + \cdots + \lambda_r e^r$$

where each λ_i is an integer. When writing down 1-chains we omit any edges which have coefficient 0. The 1-chain in which all coefficients are zero is denoted 0. We define the *sum* of the 1-chains $\Sigma \lambda_i e^i$ and $\Sigma \lambda'_i e^i$ to be

$$\Sigma \lambda_i e^i + \Sigma \lambda'_i e^i = \Sigma (\lambda_i + \lambda'_i) e^i$$

which is also a 1-chain. With this definition the set of 1-chains on G is an abelian group, denoted $C_1(G)$. (The verification that the 1-chains form an abelian group is left to the reader.)

In a similar way, a *0-chain* on G is a formal sum

$$\lambda_1 v^1 + \cdots + \lambda_m v^m$$

where $v^1, \ldots, v^m$ are the vertices of G. 0-chains are added by adding coefficients, and the set of 0-chains is an abelian group, denoted $C_0(G)$.

Example 1.18. The *1-chain associated to a path* is obtained in the manner described before 1.17. Note that the edges in a path need not be distinct (unless the path is simple), so that the final coefficient of an edge may be a number other than 0, 1 or -1. For instance, the 1-chain associated with the path $v^4 e^1 v^2 e^2 v^1 e^6 v^4 e^1 v^2 e^3 v^3$ on the graph illustrated is $e^1 + e^2 - e^6 + e^1 - e^3 = 2e^1 + e^2 - e^3 - e^6$.

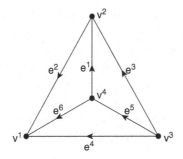

Definition 1.19. Let G be an oriented graph. For any edge $e = (vw)$ of G (regarded as a 1-chain), we define the *boundary* of e to be $\partial e = w - v$ (regarded as a 0-chain). The *boundary homomorphism*

$$\partial : C_1(G) \to C_0(G)$$

is defined by $\partial (\Sigma \lambda_i e^i) = \Sigma \lambda_i \partial (e^i)$. A *1-cycle* on G is an element $c \in C_1(G)$ such that $\partial c = 0$. The group of 1-cycles on G is denoted $Z_1(G)$. Thus $Z_1(G)$ is the kernel of the homomorphism ∂.

As an example, the 1-chain associated to any loop on G is always a 1-cycle; e.g. for the loop $v^1 e^2 v^2 e^1 v^4 e^5 v^3 e^4 v^1$ on the graph above, the associated 1-cycle is $-e^2 - e^1 - e^5 + e^4$ and

$$\partial(-e^2 - e^1 - e^5 + e^4)$$
$$= -(v^1 - v^2) - (v^2 - v^4) - (v^4 - v^3) + (v^1 - v^3) = 0.$$

Thus cycles are generalizations of loops; the following theorem describes cycles completely on a connected graph.

Now let G be a connected, oriented graph with cyclomatic number μ. Let z be a 1-cycle on G, and let $\ell^1, \ldots, \ell^\mu$ be the loops described in 1.16, for some choice of maximal tree. We denote the 1-cycle associated to ℓ^i by z^i, the direction of travel round ℓ^i being chosen so that e^i occurs in z^i with coefficient $+1$.

Theorem 1.20. *There are unique integers $\lambda_1, \ldots, \lambda_\mu$ such that*

$$z = \sum_{i=1}^{\mu} \lambda_i z^i$$

(if $\mu = 0$ this is to be read $z = 0$). In fact, λ_i is the coefficient of the edge e^i (see 1.16) in z.

Proof First suppose that G is a tree, i.e. that $\mu = 0$ and let z be a 1-cycle on G. We have to prove that $z = 0$, i.e. that there are no nontrivial 1-cycles on a tree. (This generalizes very minutely the fact that there are no loops on a tree.) The result is clear if G has no edges, so suppose it true for trees with $< k$ edges ($k \geq 1$) and let G have k edges. Let e be a non-inner edge of G, and v a free vertex of e (see 1.12(2)). The coefficient of e in z must be zero, since otherwise the coefficient of v in ∂z would not be zero. Hence z is a 1-cycle on the graph obtained by removing e and v from G. This graph is a tree with $k - 1$ edges; by induction $z = 0$.

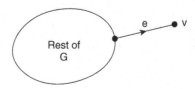

Now suppose $\mu > 0$, and let z be a 1-cycle on G. Let λ_i be the coefficient of e^i in z, with e^i as in 1.16, and consider

$$z' = z - (\lambda_1 z^1 + \cdots + \lambda_\mu z^\mu).$$

Certainly z' is a 1-cycle, since z and the z^i are. Also the coefficient of each e^i in z' is 0, by choice of the λ^i and since the coefficient of e^i in z^i is $+1$ while e^i does not occur in z^j for $j \neq i$. Thus z' is actually a 1-cycle on the graph obtained by removing the edges $e^1, \ldots, e^\mu$ from G. This graph is a tree (maximal for G) and so by the first part of the proof $z' = 0$. This proves the required formula.

It remains to verify that the given expression for z is unique. To see this, suppose that

$$\sum_{i=1}^{\mu} \lambda_i z^i = \sum_{i=1}^{\mu} \lambda'_i z^i.$$

Comparing coefficients of e^i on the two sides gives $\lambda_i = \lambda'_i$. $\square$

Remarks and Examples 1.21. (1X) Theorem 1.20 will be of considerable significance later when we come to calculate homology groups. For the present it shows that all 1-cycles, and hence all loops, on a connected graph, are in a precise sense combinations of the μ basic loops of 1.16. (It follows that all the Kirchhoff equations for loops are consequences of the equations for these basic loops.)

(2X) A *basis* for the 1-cycles on a connected graph G is a collection of 1-cycles of G in terms of which every 1-cycle on G can be expressed uniquely as a linear combination with integer coefficients. Thus $z^1, \ldots, z^\mu$ in 1.20 is a basis. But not every basis arises in this way from a maximal tree. For instance, it is impossible to find edges e^1, e^2, e^3, e^4 of the graph shown, such that the resulting cycles z^1, z^2, z^3, z^4 are the basis given by the boundaries of the four small equilateral triangles.

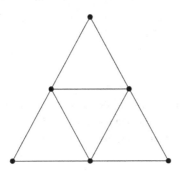

(3) 1-cycles are defined only on an oriented graph. Suppose that G is an unoriented graph and G_1 and G_2 are oriented graphs each obtained from G by

orienting the edges of G arbitrarily. Then there is a natural isomorphism

$$C_1(G_1) \to C_1(G_2)$$

defined by $(vw) \to (vw)$ or (wv) according as the edge (vw) of G receives the same or opposite orientations in G_1 and G_2. This clearly takes cycles to cycles, so that $Z_1(G_1) \cong Z_1(G_2)$. Orientation is a technical device introduced to enable the boundary homomorphism (1.19) to be defined, but the group of 1-cycles does not depend, up to isomorphism, on the orientation. In fact an equally satisfactory theory can be developed for unoriented graphs using 1-chains and 0-chains with coefficients in the group $\mathbb{Z}_2$ of integers modulo 2: this will be dealt with in a more general context in Chapter 8. The '$\mathbb{Z}_2$ theory' is, however, inadequate for some situations we shall meet later, and that is why the other theory has been introduced from the start.

(4X) Suppose that G is a connected graph with cyclomatic number μ. Suppose that there exist edges $e^1, \ldots, e^\mu$ and loops $\ell^1, \ldots, \ell^\mu$ such that, for each i, e^i is an edge of ℓ^i but not of ℓ^j for $j \ne i$. Show that removing $e^1, \ldots, e^\mu$ from G leaves a maximal tree.

(5X) A graph G is called *bipartite* if its vertex set can be partitioned into two disjoint subsets V_1, V_2 in such a way that every edge of G joins a vertex in V_1 to a vertex in V_2. Show that G is bipartite if and only if every loop in G has an even number of edges. (Hence a tree is certainly bipartite!) [Hint for 'if'. Concentrate on one component G_1 of G. Define a homomorphism $\eta : C_1(G_1) \to \mathbb{Z}_2$ by $\eta c = 0$ or 1 according as the sum of coefficients in c is even or odd. Show (using 1.20) that $\eta z = 0$ for any 1-cycle z on G_1. Now look at the 1-chains associated to paths from a fixed vertex v of G_1 to an arbitrary vertex w of G_1. A direct proof can also be given by induction on $\mu(G_1)$.]

(6XX) Let $\ell^1, \ldots, \ell^k$ be loops on an oriented connected graph and let $z^1, \ldots, z^k$ be the 1-cycles given by these loops with a choice of direction. Let us call the loops 'independent' if, for all integers $\lambda_1, \ldots, \lambda_k$, we have

$$\lambda_1 z^1 + \cdots + \lambda_k z^k = 0 \Rightarrow \lambda_1 = \cdots = \lambda_k = 0.$$

Find loops ℓ^1, ℓ^2, ℓ^3 on the complete graph with four vertices which are independent, but such that z^1, z^2, z^3 do not form a basis for the 1-cycles.

Show that two loops are independent if and only if they are distinct, and that three loops are independent if and only if none of the z^i is an integral linear combination of the other two.

(7X) Let G be an oriented graph with k vertices containing no oriented loop (an oriented loop is one in which all the edges point the same way round the loop). Show that it is possible to label the vertices of G by $v^1, \ldots, v^k$ in such a way that any oriented edge of G is of the form $(v^i v^j)$ with $i > j$. [Hint. Use induction on k. Note that, unless G has no edges, there must exist a vertex of G which is the beginning point of some edge but not the end point of any edge, for otherwise G would certainly contain an oriented loop.] The above result is used in Bernstein, Gel'fand and Ponomarev (1973) on p. 24 – note that they use 'cycle' in the sense of our word 'loop'. This paper contains deep connexions between graph theory and linear algebra.

Having investigated the kernel of the boundary homomorphism (1.19), namely the 1-cycles, we shall say something about its image.

Definition 1.22. Let G be an oriented graph. The *augmentation map* is the homomorphism

$$\varepsilon : C_0(G) \to \mathbb{Z}$$

given by

$$\varepsilon(\Sigma \lambda_i v^i) = \Sigma \lambda_i.$$

Theorem 1.23. *Let G be a connected oriented graph. Then the sequence*

$$C_1(G) \xrightarrow{\partial} C_0(G) \xrightarrow{\varepsilon} \mathbb{Z}$$

is exact, that is to say Image ∂ = Kernel ε.

Proof To show that Image $\partial \subset$ Kernel ε it is enough to verify that, for any edge e of G, $\varepsilon \partial e = 0$. This is immediate. For the converse, any element of Kernel ε can be written in the form

$$\sum_{i=1}^{m-1} \lambda_i (v^i - v^m)$$

(where $v^1, \ldots, v^m$ are the vertices of G), each λ_i being an integer. Since G is connected there is a path from v^m to v^i and the 1-chain c^i associated to this path has $\partial c^i = v^i - v^m$. Thus the above 0-chain equals

$$\partial(\Sigma \lambda_i c^i) \in \text{Image } \partial.$$

□

★ Planar graphs

A planar graph is a graph which is isomorphic to a graph in the plane $\mathbb{R}^2$. In this section we shall investigate some properties of planar graphs, mostly connected with the idea of cyclomatic number. At the end of the section we shall take up the question of whether our insistence that graphs have straight edges places any essential restriction on the class of planar graphs.

Some proofs will be omitted from this section and others will not be given in full detail; this is in order to avoid devoting undue space to what is, in fact, a subject of peripheral interest to the study of graphs in more general 'surfaces' in Chapter 9. At the end of the section the reader will need to be familiar with some point-set topology if he is to fill in all the details in the proof of 1.34. I hope, however, that the general idea of the proof will be clear to readers without such technical knowledge.

Definitions 1.24. A *polygonal arc* joining two points a and b in $\mathbb{R}^N$ is a sequence of straight segments placed end-to-end, starting at a and finishing at b. A polygonal arc is called *simple* if it has no points of self-intersection except that possibly $a = b$; thus a simple polygonal arc with $a \neq b$ can be thought of as a realization of the abstract graph given, for some $n \geq 2$, by

$$V = \{v^1, \ldots, v^n\} \quad (n \text{ distinct vertices})$$
$$E = \{\{v^1, v^2\}, \{v^2, v^3\}, \ldots, \{v^{n-1}, v^n\}\}.$$

A *simple closed polygon* is a simple polygonal arc with $a = b$; equivalently it is a realization of the abstract graph given, for some $n \geq 3$, by the same V and E as above, except that E has an additional edge $\{v^n, v^1\}$.

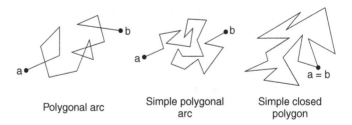

Polygonal arc   Simple polygonal arc   Simple closed polygon

It is very plausible that a tree T realized in $\mathbb{R}^2$ will not separate the plane – i.e., that any two points of the plane not on T can be joined by a polygonal arc which does not intersect T. Indeed this can be proved quite easily by induction on the number of vertices of T, and the same proof works when T is replaced by a disjoint union of trees (a 'forest'). It is also very plausible that a simple closed polygon does separate the plane, and this is the content of the next theorem.

We call a subset of the plane *bounded* if it can be contained inside a suitable circle. Otherwise we call it *unbounded*. We say that a subset P of the plane *separates* the plane into two disjoint subsets I and O if the following hold.

(i) $I \cup O = \mathbb{R}^2 \setminus P, I \cap O = \emptyset$.
(ii) Any two points of I (resp. O) can be joined by a polygonal arc, hence by a simple polygonal arc, in I (resp. O).
(iii) Any polygonal arc joining a point of I to a point of O intersects P.

(These say that $\mathbb{R}^2 \setminus p$ has precisely two 'path components'.)

1.25. Jordan Curve Theorem for Polygons *Let P be a simple closed polygon in the plane. Then P separates the plane into two disjoint subsets I and O of which precisely one, O say, is unbounded. (We call I the 'inside' and O the 'outside' of P.) Furthermore if e is any edge of P and x and y are points of $\mathbb{R}^2 \setminus P$ sufficiently close to e, on opposite sides of it and away from the end-points, then precisely one of x, y belongs to I and (hence) the other belongs to O.* □

A proof of 1.25 can be found in the book by Courant and Robbins listed in the References. The last sentence of 1.25 follows from their proof, or from the rest of the theorem together with the remarks on trees above.

A graph G in $\mathbb{R}^2$ will divide the plane into 'regions', two points not on G being in the same region if and only if they can be joined by a polygonal arc in $\mathbb{R}^2$ not intersecting G. Exactly one of these regions will be unbounded. (The graph G can be entirely contained inside a suitable circle, and any unbounded region will contain points outside this circle; on the other hand it is clear that any two points outside the circle belong to the same region.) Thus according to 1.25 a simple closed polygon will divide the plane into two regions. Our first objective is to count the regions for an arbitrary graph G.

Let e be an edge of a graph G in $\mathbb{R}^2$. Points near e and on one side of e will belong to one region, and e is called *adjacent* to that region; thus e is adjacent either to one region or to two different regions. It is clear that if e is part of a loop on G, then by 1.25 the regions on the two sides of e will be distinct; e will be adjacent to two regions. The converse is also true: if e is not part of any loop on G then e is adjacent to only one region. A proof is suggested in 1.30(3) below. Combining this with the fact, already noted in 1.9(4) for arbitrary graphs, that an edge of a graph belongs to a loop if and only if removing the edge does not increase the number of components of the graph, we have the following result.

Proposition 1.26. *Removing an edge e from a graph in $\mathbb{R}^2$ leaves the number of components of the graph unchanged if and only if it reduces the number of regions by one.*

It is now a straightforward matter to count the regions.

Theorem 1.27. *Let G be a graph in $\mathbb{R}^2$ with $\alpha_0(t)$ vertices, $\alpha_1(G)$ edges and $c(G)$ components. Then G divides the plane into $r(G) = \alpha_1(G) - \alpha_0(G) + c(G) + 1$ regions.*

Proof Let us remove the edges from G one at a time, leaving the vertices untouched, and note the effect on the value of $\alpha_1 - \alpha_0 + c + 1 - r$. In fact 1.26 says precisely that the value remains constant, and since when all the edges have gone it is $0 - \alpha_0(G) + \alpha_0(G) + 1 - 1 = 0$ it must have been 0 to begin with. This is the result. □

The formula for $r(G)$ given in 1.27 in known, at least for connected graphs ($c = 1$), as *Euler's formula*. For $c = 1$ it reads

1.28 $$\alpha_0(G) - \alpha_1(G) + r(G) = 2,$$

or equivalently

1.29 $$r(G) = \mu(G) + 1.$$

Remarks and Examples 1.30. (1X) The most noticeable thing about the formula in 1.27 is that it exists at all: the number of regions depends only on the abstract structure of G and not on the way in which G is placed in the plane. It is possible to draw 'graphs' in surfaces other than the plane, for example in the torus surface. The edges will be curved lines on the surface, and again we can count regions. For the graph illustrated below, $\alpha_0 = 6, \alpha_1 = 7, c = 1$ and $r = 2$. However it is possible to find isomorphic graphs in the torus (these will have the same α_0, α_1 and c), for which $r = 1$ or $r = 3$. Thus the number of regions does not depend only on the abstract structure of the graph. This phenomenon will be investigated in Chapter 9. What goes wrong when the argument leading to 1.26 is applied to graphs in the torus?

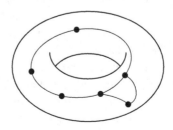

Planar graphs

(2X) On the circumference of a circle in the plane n points are chosen and the chords joining every pair of points drawn. Assuming that no three of these chords are concurrent (except at one of the n points) show that the interior of the circle is divided into $1 + \binom{n}{2} + \binom{n}{4}$ regions. [Hint. Use Euler's formula. The number of points of intersection of chords inside the circle is $\binom{n}{4}$ since any four of the n points uniquely determine such a point of intersection. It is amusing to note that for $n = 1, 2, 3, 4, 5, 6$ the number of regions is $1, 2, 4, 8, 16, 31$.]

(3X) Here is a suggestion for proving that, if an edge e of a graph G in $\mathbb{R}^2$ is not part of any loop on G, then the regions on the two sides of e are the same. First dispose of the case when G is a tree (e.g. using the remarks preceding 1.25), and then proceed by induction on the number of edges in G. Remove an edge e' from a loop ℓ on G, noting that $e \neq e'$ and e, apart possibly from its end-points, is entirely inside or entirely outside ℓ. Choose a polygonal arc from one side of e to the other, using the induction hypothesis. The only snag is that this arc may cross e'. It is not difficult to change the arc so that it doesn't cross e'; a hint is given in the diagram.

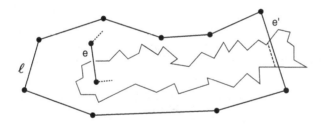

We shall now look more closely at the formula 1.29 for connected graphs in $\mathbb{R}^2$. We already know, from 1.20, that there is a basis for the 1-cycles on G

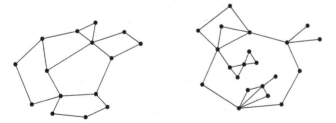

which consists of $\mu(G)$ elements. It is a consequence of a theorem of group theory (see A.30) that every basis has this number of elements. Now in the case of the left-hand illustration above (where $\mu(G) = 6$) it is very plausible that the boundaries of the six bounded regions give six cycles forming a basis for the 1-cycles on G. (It is at least 'clear' that every loop on G can be obtained by combining these six loops suitably.) Perhaps this is not so obvious in the case of the right-hand illustration simply because it is not so clear which are the loops to take. If we can associate one basic loop with each bounded region, then this will confirm that $r(G) = \mu(G) + 1$, the '1' on the right-hand side coming from the unbounded region. In fact such an association is possible, as stated in the next proposition, which will be given without proof.

Let G be a connected oriented graph in $\mathbb{R}^2$, and let R be one of the regions into which G divides $\mathbb{R}^2$. The *frontier* $\mathscr{F}(R)$ of R consists of those edges of G which are adjacent to R, and their end-points. Thus $\mathscr{F}(R)$ is a subgraph of G. Now suppose that R is a bounded region, and consider just the graph $\mathscr{F}(R)$ in $\mathbb{R}^2$, deleting the remainder of G.

Proposition 1.31. *The edges of $\mathscr{F}(R)$ which are adjacent to the unbounded region (when the remainder of G is deleted), together with their end-points, form a simple closed polygon. This gives us a loop on G; the loops associated in this way to the various bounded regions give a basis for the 1-cycles on G.*

□

Note that the basis provided in this way may be different from any given by a maximal tree; see 1.21(2). As an example of 1.31, the frontier of one of the regions in the right-hand illustration above is drawn below, with the associated loop drawn heavily.

Another noteworthy point about the basis described in 1.31 is that any edge of G belongs to at most two of the basic loops. A remarkable theorem of S. Mac Lane (1937) asserts that the converse is true: if a connected graph G possesses loops $a^1, \ldots, a^\mu$, whose corresponding 1-cycles form a basis for the 1-cycles on G and which have the property that each edge of G belongs to at most two of the loops a^i, then G is planar. Two other criteria for the planarity of a graph are known; one is due to Kuratowski and the other to Whitney (1932). See the books by Gross & Yellen, Diestel and Prasolov (*Combinatorial and Differential Topology*) listed in the References. There is also a proof of

Kuratowski's theorem which simultaneously proves Fáry's theorem (below) in Thomassen (1981).

We can make a further deduction from 1.31. Suppose that G (in $\mathbb{R}^2$) is connected, with $\mu(G) \geq 1$. Each loop associated to a bounded region will have at least three edges, so that the sum of the numbers of edges in these loops is at least 3μ. Also the frontier of the unbounded region will contain at least three edges, and adding these in gives a sum of at least $3\mu + 3$. This sum is $\leq 2\alpha_1(G)$ since an edge of G belongs to at most two of the loops and to at most one if the edge is adjacent to the unbounded region. From this we deduce the following.

Proposition 1.32. *If G is a connected graph in $\mathbb{R}^2$ satisfying $\mu(G) \geq 1$, then $3\alpha_0(G) - \alpha_1(G) \geq 6$.* □

The case $\mu(G) = 0$ is left to the reader. As an example of the proposition let G be the complete graph with five vertices. For this G, $\alpha_0 = 5$, $\alpha_1 = 10$. (Also $\mu = 6 \geq 1$.) Thus $3\alpha_0 - \alpha_1 < 6$, and G is not planar.

Example 1.33. (X) Show that if G is a connected bipartite graph (1.21(5)) in $\mathbb{R}^2$ satisfying $\mu(G) \geq 1$, then $2\alpha_0(G) - \alpha_1(G) \geq 4$. Find a bipartite graph which is not planar.

Given a graph G we can 'subdivide' it by adding extra vertices at interior points of edges; this automatically increases the number of edges by the same amount, thus leaving $\mu(G)$ unchanged. For example if we subdivide the complete graph with five vertices by adding a single vertex on one of the edges the new graph has $\alpha_0 = 6$, $\alpha_1 = 11$ and it is no longer true that $3\alpha_0 - \alpha_1 < 6$. Thus 1.32 does not show that the subdivided graph is non-planar. This indicates, of course, how feeble the result 1.32 is; all the same it is a priori conceivable that by adding enough extra vertices on the edges of a non-planar graph we could make it planar – this is rather like trying to draw it in the plane with curved edges, since these could be approximated by broken lines.

A theorem of K. Wagner (1936) and I. Fáry (1948), known as Fáry's Theorem, implies that this does not happen; if some subdivision of a graph G is planar, then G is planar. It would take us too far out of our way to give a full and rigorous proof of Fáry's theorem, since this would involve the formal introduction of some notions of point-set topology with which we are not otherwise concerned. However there follows a proof that is complete except for these details. It is a variant of Fáry's original proof, and was found by P. Damphousse. In fact Damphousse's argument proves a stronger result than that proved by Fáry (this stronger result is stated in Fáry's paper). (See also the books of Prasolov (*Combinatorial and Differential Topology*) and Gross & Yellen listed in the References; also Bryant (1989).)

We shall consider 'curved graphs' in the plane which differ from the graphs considered so far only in that their edges are allowed to be curved lines (precisely: homeomorphic images of the closed interval [0, 1]) instead of straight segments. Thus a *curved graph* $\mathscr{G}$ consists of finite set of points (vertices) in the plane, some pairs of distinct vertices being joined by curved lines (edges) which do not intersect except possibly at end-points. An edge does not intersect itself either: it is a simple curve. Furthermore any particular pair of vertices is joined by at most one edge. It is possible to formulate and prove Euler's formula 1.27 for curved graphs: both formulation and proof require the most general form of the Jordan Curve Theorem which states, roughly speaking, that a simple closed curve always separates the plane into two regions. (See the books of Prasolov (*Combinatorial and Differential Topology*) and Wall listed in the References.) In this sketch we shall assume the validity of Euler's formula for curved graphs, and of some other 'plausible' results to be mentioned later.

The weak form of Fáry's theorem asserts that the abstract graph underlying $\mathscr{G}$ can be realized in $\mathbb{R}^2$ (see 1.3), i.e. that there is a (straight) graph G in $\mathbb{R}^2$ with the same abstract structure as $\mathscr{G}$. The strong form is as follows.

1.34. Fáry's Theorem on straightening planar graphs *With the above notation, there is a homeomorphism of the plane (that is, a bijective continuous map $\mathbb{R}^2 \to \mathbb{R}^2$ with continuous inverse) mapping $\mathscr{G}$ onto a (straight) graph G in such a way that vertices of $\mathscr{G}$ are taken to vertices of G and edges to edges.*

Proof The proof is in several parts. First we make changes to $\mathscr{G}$ to simplify its structure (I); (II)–(IV) are consequence of these changes. (V) is the central argument (by induction on the number of vertices of $\mathscr{G}$) and (VI) is a small result needed in (V). We assume throughout that $\mathscr{G}$ has at least four vertices, the other cases being easily disposed of.

(I) We turn $\mathscr{G}$ into a *triangular* graph $\mathscr{G}^\Delta$, that is a curved graph for which each one of the regions into which $\mathscr{G}^\Delta$ divides the plane has frontier (see before 1.31) a simple closed curve consisting of precisely three edges of $\mathscr{G}^\Delta$. This is achieved by a sequence of operations on $\mathscr{G}$, as follows. Select any two vertices of $\mathscr{G}$ which are on the frontier of the same region and which are not already connected by any edge of $\mathscr{G}$; connect them by a simple curve within the region to give a graph $\mathscr{G}'$. Now select a pair of vertices of $\mathscr{G}'$ which are on the frontier of the same region defined by $\mathscr{G}'$ and are not connected by any edge of $\mathscr{G}'$ and connect them within the region. This procedure is continued until it is no longer possible to find a pair of vertices of the sort described – this must eventually happen since no new vertices are introduced and consequently the number of possible edges is bounded. An example is drawn below; the dotted edges are the ones introduced.

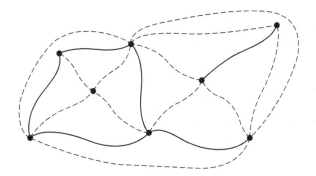

To see that the result is triangular, suppose that a region has frontier consisting of say four edges, as in the diagram below. By the construction A and C must be joined by an edge outside the region and so must B and D. But this is impossible in $\mathbb{R}^2$. (This is 'obvious' but formally requires the Jordan Curve Theorem.)

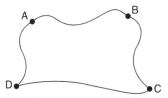

(II) In any triangular graph with at least four vertices there can be no vertex of order 0, 1 or 2. To see, for example, that there can be no vertex of order 2 suppose that there is one – B say, in the diagram above. As there is at least one vertex other than A, B, C there must be such a vertex on the frontier of one of the regions adjacent to the arc ABC. But this vertex is not joined to B, so (I) has not been completed.

(III) In any triangular graph with at least four vertices there are at least four vertices of order ≤ 5. From the triangularity it follows that $\mathscr{G}$ is connected and $3r = 2\alpha_1$ (r being the number of regions) and this together with Euler's formula 1.28 shows $6\alpha_0 - 2\alpha_1 = 12$. Now writing N_i for the number of vertices of order i (so that $N_0 = N_1 = N_2 = 0$ from (II)), we have

$$\alpha_0 = N_3 + N_4 + N_5 + \cdots \text{ and } 2\alpha_1 = 3N_3 + 4N_4 + 5N_5 + \cdots$$

so that $12 = 6\alpha_0 - 2\alpha_1 = 3N_3 + 2N_4 + N_5 - N_7 - 2N_8 - \cdots \leq 3N_3 + 2N_4 + N_5$.

Hence $N_3 + N_4 + N_5 \geq \frac{1}{3}(3N_3 + 2N_4 + N_5) \geq 4$, as required.

(IV) There is at least one vertex v of $\mathscr{G}^\Delta$, of order ≤ 5 and not on the frontier of the unbounded region. This is because there are precisely three vertices of

$\mathscr{G}^\Delta$ on the frontier of the unbounded region, and by (III) at least four vertices of $\mathscr{G}^\Delta$ of order ≤ 5. Now consider the edges with v as one end-point; they can be ordered, say anticlockwise, round v. (This statement is not as innocent as it looks, for the edges could for example spiral round v in a complicated manner. What they cannot do is cross each other, and that is what makes the result, which follows from the Jordan Curve Theorem, true.) Consecutive pairs of these edges will terminate in pairs of vertices, and these pairs will be joined by edges of $\mathscr{G}^\Delta$ which will form a simple closed curve $\mathscr{C}$ enclosing v.

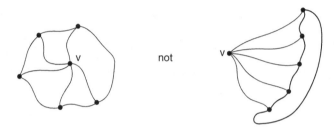

(It is not hard to see that the right-hand picture can only arise when v is on the frontier of the unbounded region.)

(V) This is the induction argument by which the theorem is proved. Suppose 1.34 is true whenever $\mathscr{G}$ has $< k$ vertices ($k \geq 4$) and let $\mathscr{G}$ have k vertices. (I)–(IV) apply to $\mathscr{G}$, which turns into a triangular graph $\mathscr{G}^\Delta$; we then remove the v of (IV) together with its attendant edges, leaving a graph $\mathscr{G}_1$, say. Now use the induction hypothesis to choose a homeomorphism h of the plane to itself mapping $\mathscr{G}_1$ on to a straight graph G_1. This will take $\mathscr{C}$ (see (IV)) to a simple closed polygon C say, with 3, 4 or 5 sides and with no vertex inside it since h will preserve the interior of a simple closed curve. It remains to 'fill in' the missing vertex and edges of $\mathscr{G}$ and change h inside $\mathscr{C}$ so that it maps them to a point P inside C and straight segments joining P to the vertices around C. Thus we need, in the first place, the existence of point P within C from which all vertices of C are 'visible'. This is dealt with in (VI).

Having found such a P it is possible to redefine h inside $\mathscr{C}$ as above. (First we ensure that $h(v) = P$, then that the curved edges through v are mapped to the straight edges through P. It then remains to extend h to the interiors of the 'curved triangles' inside $\mathscr{C}$. This is a standard procedure in point-set topology and we shall not go into the details.) Since $\mathscr{G}$ is a subgraph of $\mathscr{G}^\Delta$, this completes the induction, for h will automatically take $\mathscr{G}$ on to a straight graph.

(VI) Within any simple closed polygon C with 3, 4 or 5 sides there is always a point P from which all the vertices of C are visible. The proof of this is left as an exercise, with the following hint for a pentagon. Divide the interior of C

into three triangles by two straight segments connecting vertices of C and lying entirely inside C. (It can be proved, in fact, that the interior of any simple closed polygon with n sides can be divided into $n-2$ triangles by drawing $n-3$ such segments, intersecting only at vertices of the polygon.) Since there are three triangles one of them must be 'central' in the sense that it is joined to each of the other two along an edge. In the diagrams the central triangle is shaded.

 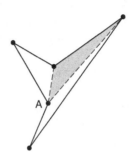

It is not difficult to find points within the central triangle close to the vertex A from which all vertices of the pentagon are visible.

Note that pentagons are a critical case: there exist hexagons within which there is no point from which all six vertices are visible.

This completes the sketch of a proof of Fáry's Theorem. ★

★ Appendix on Kirchhoff's equations

In this appendix we shall present a proof of the fact that Kirchhoff's equations (see the section starting after 1.7) determine the currents uniquely for any circuit. Consider the basic loops $\ell^1, \ldots \ell^\mu$ described in 1.16. The $(K1)$ equation of the loop ℓ^i will be written $L_i = b_i$ where L_i is the part involving the (unknown) currents and b_i is the total e.m.f. in the loop. As has already been noted in 1.21(1) the $(K1)$ equation of any loop is a consequence of these μ equations; it was also noted during the earlier discussion that all but one of the $(K2)$ equations are independent. Writing $V_i = 0$ for the equation of the vertex v^i we have the following, which establishes the required result.

Proposition 1.35. *The $\alpha = \alpha_1$ equations*

$$L_1 = b_1, \ldots, L_\mu = b_\mu, \quad V_2 = 0, \ldots, V_n = 0 \ (n = \alpha_0)$$

are independent.

Proof Suppose that

$$\lambda_1 L_1 + \cdots + \lambda_\mu L_\mu = \nu_2 V_2 + \cdots + \nu_n V_n. \qquad (*)$$

We prove that $\lambda_1 = \cdots = \lambda_\mu = 0$ which, in view of the independence of $V_2, \ldots, V_n$, implies the result.

We introduce some new notation at this point. It is not strictly necessary to the argument, but allows a more succinct presentation. The ideas of 'cochain' and 'coboundary' are of great importance in algebraic topology, but we shall not have occasion to make use of them apart from in this one case.

In order to allow the resistances in the wires to be real numbers and not just integers, it is convenient to work with 'real chains', that is linear combinations of vertices or edges in which the coefficients are real numbers. In 1.17 the λ_i become elements of $\mathbb{R}$, the real numbers, and $C_0(G)$ and $C_1(G)$ become real vector spaces. Likewise in 1.19 the boundary homomorphism ∂ becomes a linear map. We shall keep to the old notation, but, just for this proof, work over $\mathbb{R}$ instead of $\mathbb{Z}$.

A *0-cochain* is a linear map $\gamma^0 : C_0(G) \to \mathbb{R}$ and a *1-cochain* is a linear map $\gamma^1 : C_1(G) \to \mathbb{R}$. The value of a cochain γ on a chain c is denoted $\langle \gamma, c \rangle \in \mathbb{R}$, and the i-cochains form a real vector space $C^i(G)$ ($i = 0, 1$) with addition and scalar multiplication defined by

$$\langle \gamma + \gamma', c \rangle = \langle \gamma, c \rangle + \langle \gamma' c \rangle; \quad \langle \lambda \gamma, c \rangle = \lambda \langle \gamma, c \rangle.$$

There is a *coboundary homomorphism*

$$\delta : C^0(G) \to C^1(G)$$

defined by

$$\langle \delta \gamma^0, c_1 \rangle = \langle \gamma^0, \partial c_1 \rangle.$$

Now writing x_j for the (unknown) current in the oriented edge e^j and r_j for the resistance of this (unoriented) edge, L_i will have the form

$$L_i = \sum_{j=1}^{\alpha} \epsilon_{ij} r_j x_j \quad (\epsilon_{ij} = -1, 0 \text{ or } 1).$$

Taking the direction round the loop ℓ^i to be given by the orientation of the edge e^i, it follows that $\epsilon_{ij} = 0$ if $i \neq j$ and $j \leq \mu$, while $\epsilon_{ii} = +1$. This is because e^j

does not occur in any of the loops ℓ^i for $i \neq j \leq \mu$. Hence the $\mu \times \alpha$ matrix $A = (\epsilon_{ij})$ has the $\mu \times \mu$ identity matrix as its first μ columns. This will be of significance later.

With L_i we associate the 1-cochain, also called L_i, which takes the edge e^j to $\epsilon_{ij} r_j \in \mathbb{R}$. In the same way we can associate a 1-cochain with V_i, and it is straightforward to check that $V_i = \delta v^i$, where v^i is the 0-cochain taking the value 1 on the vertex v^i and 0 on every other vertex. Thus the equation $(*)$ above says precisely that $\lambda_1 L_1 + \cdots + \lambda_\mu L_\mu$ is in the image of δ.

It follows from the formula for δ that if we evaluate $\lambda_1 L_1 + \cdots + \lambda_\mu L_\mu$ on any cycle the result will be zero. Let us evaluate it on the cycles $z^1, \ldots, z^\mu$ given by the basic loops $\ell^1, \ldots, \ell^\mu$. The result is μ linear equations in $\lambda_1, \ldots, \lambda_\mu$ and a little calculation reveals that the coefficient matrix is $(\langle L_i, z^j \rangle)$. With A as above and $R = \mathrm{diag}(r_1, \ldots, r_\alpha)$ this coefficient matrix is precisely $ARA^\top$ ($\top$ for transpose). We shall show that this matrix is nonsingular.

Now each resistance is strictly positive, so we can define a nonsingular matrix $S = \mathrm{diag}(\sqrt{r_1}, \ldots, \sqrt{r_\alpha})$. Clearly $ARA^\top = (AS)(AS)^\top$. However A has rank μ since it contains the $\mu \times \mu$ identity matrix. Hence AS has rank μ. It is now a standard fact of linear algebra that, AS being real, the product with its transpose is nonsingular. (One way to prove this is to note that $(AS)(AS)^\top$, being symmetric, is congruent to a diagonal matrix. This is a standard result of linear algebra.) Thus $\lambda_1 = \cdots = \lambda_\mu = 0$, as required. $\square$

It is not difficult to work this argument backwards and show that, if $ARA^\top$ is singular, then the Kirchhoff equations are dependent or contradictory. (The backward argument uses the fact that a cochain which vanishes on every cycle is necessarily a coboundary. This holds over $\mathbb{R}$, but not over $\mathbb{Z}$.) Thus the nonsingularity of $ARA^\top$ is necessary and sufficient for the Kirchhoff equations to determine the currents uniquely. It is a curious fact that this condition is much weaker than the condition '$r_i > 0$ for all i' which was used at the end of the proof. For example when $\mu = 1$ it says that the sum of the resistances in the unique loop should be non-zero.

In Kirchhoff's original 1847 paper, he shows how to determine the currents (in the physical situation $r_i > 0$) and also shows that all the $(K1)$ equations depend on the μ equations for the basic loops. Parts of this paper are readily available in translation in the book of Biggs, Lloyd & Wilson listed in the References. Other useful references are the books of Bollobás, Wilson & Beinecke and Tutte. ★

2
Closed surfaces

A surface is usually thought of as something smooth and rounded, like the surface of a sphere or a torus. In order to study surfaces systematically we shall, as described in the Introduction, assume that they are divided up into triangles.[†] It is therefore necessary at the outset to pinpoint the decisive property which an object must possess in order to be called a surface, and to interpret this as a property of the triangulation.

The decisive property is this. Take a point P on the surface; then the points of the surface close to P make up a little patch on the surface, this patch being essentially a two-dimensional disk. The disk may be bent or stretched, but must not be torn (that is, 'essentially' means, in technical language, 'homeomorphic to'). In terms of triangulations this means that the edges opposite P in the triangles with P as a vertex form a simple closed polygon. A definition along

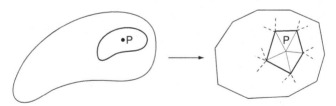

these lines is made precise in 2.1. Note that the definition excludes surfaces like the Möbius band and cylinder which have boundary edges or rims; to emphasize that our surfaces have no boundary we call them 'closed surfaces'. The more general concept of 'surface with boundary' is explained and used in Chapter 9.

[†] This is no restriction for compact surfaces (e.g. surfaces which are closed subsets of a bounded region of euclidean space). Indeed with the more general concept of infinite triangulation a wider class of surfaces can be triangulated. These matters are dealt with in detail in Chapter I of the book by Ahlfors and Sario listed in the References and in the book *Elements of Combinatorial and Differential Topology* by Prasolov, also listed. There is also a proof in Doyle & Moran (1968).

In the formal definition of closed surface (2.1) the triangles will be ordinary flat triangles with straight edges. This is simply because flat triangles are easier to define and to work with than curvy ones. The reader should not feel uneasy because the smooth rounded surface has become a somewhat angular and kinky object; he is quite at liberty to continue drawing rounded surfaces – and indeed the author will exercise that liberty too. In Chapter 1 we saw that no generality was lost by specifying that the edges of a graph should be straight. In a similar way every surface with curved triangles can be realized (at any rate in $\mathbb{R}^5$) with flat triangles, but the question of realization is not taken up until 3.19.

The first objective of this chapter is to construct all closed surfaces and to exhibit them in a convenient way by means of a 'polygonal representation'. From this representation it is possible, by a sequence of 'scissors and glue' operations to put every closed surface into exactly one of a list of standard forms – this list appears in 2.8. We can then go on to derive a geometrical description of closed surfaces in terms of three basic ingredients – sphere, torus and projective plane – and one mixing operation – connected sum. In this way we can picture what all closed surfaces 'look like'.

Finally we show how to distinguish different closed surfaces from one another, and in particular how to prove that no two closed surfaces in the list 2.8 are the same.

There is an intrinsic interest in trying to describe all possible surfaces, but the longer-term purpose so far as this book is concerned is to provide plenty of good and geometrical examples with which to illustrate later theorems.

Closed surfaces and orientability

A graph in which there are no isolated vertices consists of a collection of closed segments (edges) which satisfy an intersection condition, namely that two edges intersect, if at all, in a common end-point (a vertex). We now introduce a two-dimensional building block and start to make two-dimensional objects, again restricting the possibilities by imposing an intersection condition. The new building blocks are triangles, which may be any shape or size. Note that any triangle comes equipped with three vertices and three straight edges. The *intersection condition* we impose is this:

Two triangles either

(i) are disjoint or
(ii) have one vertex in common or

(iii) have two vertices, and consequently the entire edge joining them, in common.

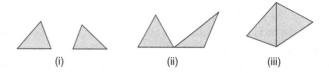

Some configurations which are not allowed are pictured below:

(In the terminology of Chapter 3, a collection of triangles satisfying the intersection condition is said to form a pure two-dimensional simplicial complex.)

As in the case of graphs, we have the concept of connectedness: a collection M of triangles satisfying the intersection condition is called *connected* if there is a path along the edges of the triangles from any vertex to any other vertex. Indeed the set of edges and vertices of triangles in M forms a graph M^1 and we have just defined 'M is connected' to mean 'M^1 is connected'.

There is one other concept needed in order to define closed surfaces. Consider a vertex v of some triangle of a collection satisfying the intersection condition. The edges opposite v in the triangles of M having v as a vertex will form a graph, called the *link* of v.

Definition 2.1. A *closed surface* is a collection M of triangles (in some euclidean space) such that

(i) M satisfies the intersection condition (see above)
(ii) M is connected
(iii) for every vertex v of a triangle of M, the link of v is a simple closed polygon.

Remarks and Examples 2.2. (1) A hollow tetrahedron is a closed surface, the link of any vertex being the three edges of the opposite face. However two hollow tetrahedra with a common vertex do not form a closed surface since the link of the common vertex is two simple closed polygons and not one.

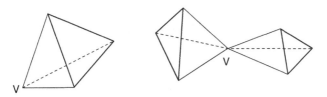

(2) It is not difficult to show that any edge of a closed surface M is on exactly two triangles of M. (Consider the link of one of the end-points of the edge.) Note that two hollow tetrahedra with a vertex in common have this latter property, but, as pointed out in (1), do not form a closed surface. Two hollow tetrahedra with an edge in common do not satisfy even the weaker condition that every edge is on exactly two triangles.

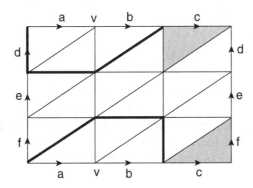

(3) The diagram above shows a schematic representation of a closed surface containing 18 triangles. The pairs of similarly marked edges are to be 'identified' with arrows corresponding so that, for example, the two shaded triangles actually have the edge c in common. The link of any vertex is then a simple closed polygon – the link of v is drawn heavily in the diagram, the two halves fitting together to form a simple closed hexagon. The closed surface so described is called *a triangulation of the torus* (sometimes abbreviated to 'torus') since, informally, if we stick together the similarly marked edges in $\mathbb{R}^3$, the result is a torus.

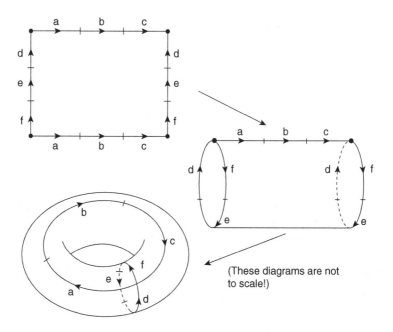

(These diagrams are not to scale!)

(4) The left-hand diagram below represents a closed surface which is a triangulation of the '*Klein bottle*' and the right-hand diagram represents a triangulation of the '*projective plane*'. (The shading in these diagrams is not relevant until later.) Neither of these surfaces can be constructed in three-dimensional space (see 9.18). A picture of the Klein bottle, in which there is a self-intersection along a circle, appears on the next page. For similar pictures of the projective plane, see 2.7(3); see also the book of Hilbert & Cohn-Vossen listed in the References.

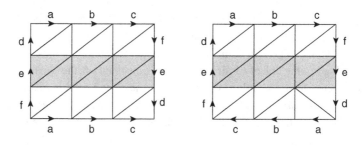

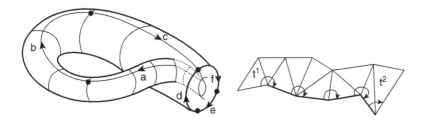

(5X) Let t^1 and t^2 be two triangles of a closed surface M. Join some vertex of t^1 to some vertex of t^2 by a simple path along the edges of M. Using the fact that the link of each vertex along this path is a simple closed polygon it is possible to construct a sequence of triangles connecting t^1 and t^2, any two consecutive triangles in the sequence having an edge in common. The arrows in the right-hand diagram above show the way in which a sequence of triangles is constructed using in turn the links of vertices on the path. Note that such a sequence cannot always be constructed on the two tetrahedra with a common vertex, but it can on the two tetrahedra with a common edge.

(6) The definition of closed surface does not allow the collection of triangles in the upper diagram below, since the link of the vertex v is not a simple closed polygon (it is a polygonal arc). The diagram shows a triangulation of the *Möbius band*, drawn underneath, which is obtained by sticking together the ends of a rectangle after a twist of 180°. The Möbius band is a surface in a more general sense which will be explained in 9.8. Reversing one of the arrows in the upper diagram we obtain a *cylinder*, in which the ends of the rectangle are identified without a twist.

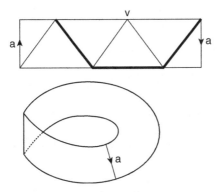

The major division amongst closed surfaces is into orientable and non-orientable closed surfaces. Once again we shall meet the concept of orientation more formally in the next chapter; for the present we shall use the following informal idea. Suppose that on each triangle of a closed surface M we draw a circular arrow (called an *orientation*) as in the diagrams.

In the left-hand diagram, the two triangles with a common edge are said to be *coherently oriented*, while in the right-hand diagram the two triangles are not coherently oriented.

Definition 2.3. A closed surface M is called (*coherently*) *orientable* if all the triangles of M can be given an orientation in such a way that two triangles with a common edge are always oriented coherently. Otherwise M is called *non-orientable*.

Example 2.4. (X) The torus in 2.2(3) is orientable, but the Klein bottle and projective plane in 2.2(4) are non-orientable. Note that the latter two closed surfaces both contain Möbius bands (see 2.2(6) and the shading in 2.2(4)). In fact it will become clear that a closed surface is non-orientable if and only if it contains a Möbius band. In the case of the Klein bottle, removing the shaded triangles in 2.2(4) leaves another Möbius band, as can be seen by identifying the two a's, b's and c's. On the other hand the diagram below illustrates removal

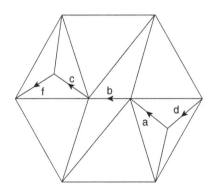

of the shaded triangles from the projective plane in 2.2(4) (allowing a little distortion of the triangles). This certainly does not contain any more Möbius bands. Thus the Klein bottle is 'more non-orientable' than the projective plane. In fact it is just twice as non-orientable, as we shall see following 2.9.

Polygonal representation of a closed surface

Let us start with a closed surface M, orient the edges arbitrarily and then label the triangles $t^1, \ldots, t^n$ and their edges $e^1, \ldots, e^m$ say. (In fact $3n = 2m$; in particular n is even.) The information needed to construct M comprises the n triangles and, for each triangle, the three oriented edges which belong to it. For example the tetrahedron on the left can be constructed from the four triangles on the right.

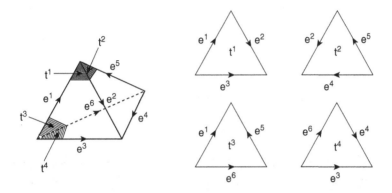

Our immediate object is to produce a more convenient plane representation of M than the collection of n disjoint triangles with labelled edges. This is achieved, in the case of the tetrahedron above, by partially assembling M in the plane: we could stick t^2 to t^1 along e^2, then t^3 to t^1 along e^1, then t^4 to t^1 along e^3.

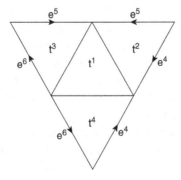

The result is a diagram which completely describes the way in which the triangles of M intersect one another. Other examples of such a plane representation are in 2.2(3), (4) and (6) above.

The same procedure can be followed for any surface M. Again we start with n disjoint triangles with labelled edges and assemble them one at a time. Triangles which have been assembled at a given stage will be called 'used' and the remainder 'unused'. It is essential to identify at each stage an edge of an unused triangle to an edge of a used one; the boundary of the plane region covered by the used triangles will then always be a simple closed polygon. There is one slight snag: we may find that the unused triangle which is added at some stage overlaps the used triangles, as in the diagram, where the shaded

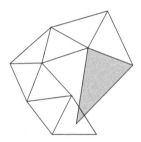

triangle was the last to be added. To overcome this we must abandon hope of having the triangles in the plane actually congruent to the triangles of M. Instead, as each triangle is used, we distort it (keeping the edges straight) so that the triangles in the plane do not overlap – for example by making the boundary polygon given by the used triangles always convex. The final representation of M will still faithfully record all the intersections of triangles of M (that is, it determines M 'up to isomorphism' in the language of Chapter 3). It remains to prove that, until all triangles have been used, there is always an unused triangle with an edge in common with the boundary polygon given by the used triangles. There is certainly an unused triangle with a vertex v in common with some used triangle, and the construction ensures that all vertices are on the boundary polygon. It is now straightforward to verify, using the fact that the link of v in M is a simple closed polygon, that there is an unused triangle with v and another vertex (and hence an edge) in common with a used triangle. (Note that the construction would break down at some stage if we started with two tetrahedra with a common vertex.)

Continuing until all the triangles of M are used, we obtain a *polygonal representation* of M as the region bounded by a simple closed polygon. This region is broken up into an even number n of triangles which correspond to the triangles of M and the boundary polygon has $n+2$ edges which occur in equally labelled pairs. The identifications necessary to recover M can be described by a *symbol*: start at any vertex and read round the boundary polygon, writing an edge or its 'inverse' according as its orientation goes with or against the direction of travel. A symbol for the tetrahedron above is

$$(e^5)^{-1}e^4(e^4)^{-1}(e^6)^{-1}e^6e^5.$$

Note that the torus in 2.2(3) could not arise directly from from this construction since it has vertices not on the boundary polygon. What happens between the above construction and such a figure as 2.2(3) or 2.2(4) is that two edges with equal labels and orientation, and having a common vertex, become identified, creating an interior vertex. This process will form part of the reduction to standard form, below. The torus in 2.2(3) has a symbol $abcd^{-1}e^{-1}f^{-1}c^{-1}b^{-1}a^{-1}fed$, while the Klein bottle in 2.2(4) has a symbol $abcfedc^{-1}b^{-1}a^{-1}fed$. Note that a symbol is essentially cyclic: we could start reading it off at a different place; also read round the other way.

The triangles of a polygonal representation of a closed surface M can be oriented coherently (all clockwise, say, but the sceptical reader can prove it by induction on the number of triangles); whether this gives a coherent orientation of M itself depends on the disposition of arrows round the boundary polygon. If the symbol contains ... a ... a^{-1} ... then the orientation will be coherent across the edges making up a; if it contains ... b ... b ... then the orientation will fail to be coherent across the edges making up b. In fact, we have the following result.

Proposition 2.5. *Suppose that a closed surface M is represented by a symbol, as above. Then M is orientable if and only if, for every letter occurring in the symbol, its inverse also occurs.* □

★ A note on realizations

Suppose we are given a simple closed polygon with an even number of sides, the plane region bounded by the polygon being divided into triangles in such a way that the sides of the polygon are edges of triangles and two triangles intersect, if at all, in a vertex or an edge. Now mark the sides of the polygon with letters and arrows, each letter being used exactly twice as in the examples of 2.2(3), (4). The diagram can be interpreted as a 'schema' for sticking triangles together, the similarly marked edges being identified with arrows corresponding. When is it possible to realize this schema by a closed surface? The problem of realizing 'schemas' of this sort will be treated in more detail in Chapter 3 (see in particular 3.20); for the present we shall be content with stating the following result. The identification of edges entails identification of certain vertices of the boundary polygon, and with this in mind we shall label the vertices of the diagram, those to be identified being given the same label, as in the diagrams below.

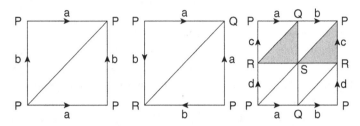

Note that none of these three determines a closed surface. For the left-hand and centre diagrams the two triangles would have all three edges in common, while for the right-hand diagram the two shaded triangles would have two vertices but no edge in common (see the intersection condition preceding 2.1).

Proposition 2.6. *A schema of the kind described above is realizable as a closed surface if and only if*

(i) *the end-points of any edge are labelled differently, and*
(ii) *no two edges (unless they are to be identified) have end-points labelled the same, and*
(iii) *no two triangles have vertices labelled the same.* □

The diagram is then called a *polygonal representation* of the closed surface.

The polygonal representations described before are a special case of these; the identifications necessary to recover M can again be described by a symbol and the orientability criterion 2.5 still holds.

Note that (i)–(iii) imply that the link of any vertex will be a simple closed polygon. The reader may care to verify this, and also that the only case in which (i) and (ii) hold but (iii) does not is the centre diagram above.

It is worth noting that any diagram of the kind described before 2.6 can be subdivided into one which represents a closed surface: subdivide every triangle into six by drawing the medians, and then subdivide every small triangle into six again by drawing the medians. One such subdivision suffices for the centre and right-hand diagrams above, but both are needed for the left-hand diagram. The subdivision obtained by drawing the medians of all triangles is called *barycentric subdivision* and will be encountered again in Chapter 5. ★

Transformation of closed surfaces to standard form

We shall describe a method, traditionally known as 'scissors and glue', of reducing any polygonal representation of a surface to precisely one of a list of standard forms. During the course of this it may be necessary – or at any rate convenient – to subdivide M barycentrically (that is, drawing the medians of all triangles) a number of times; this is in order to ensure that our scissors always cut along edges of triangles of M. Thus the final symbol will represent either M or some repeated barycentric subdivision of M. Note that if M is a closed surface and M' is obtained from M by subdividing barycentrically then M' is still a closed surface. In the diagram, the link of the central vertex contains six edges of M (drawn heavily) and twelve edges of M'.

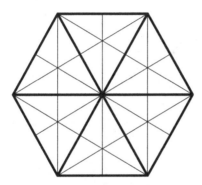

The reduction to standard form is in several steps, numbered as the parts of 2.7. The reader who turns straight to the answer (2.8) will not find himself much penalized in the sequel, since we shall not often make use of the precise nature of the steps used in the reduction to standard form; the main exception is the proof of 2.14.

2.7. The reduction algorithm

(1) First, a piece of notation. If a succession of oriented edges occurs twice on the boundary polygon, either in the same or opposite directions, then we shall always denote the succession of edges by a single letter. For example the diagram of 2.2(3) has symbol $xyx^{-1}y^{-1}$ where $x = abc$, $y = d^{-1}e^{-1}f^{-1}$. In what follows both x and x^{-1} (for example) are called *letters*. Note that, as in group theory, $(abc)^{-1} = c^{-1}b^{-1}a^{-1}$.

(2) There is a further analogy with group theory in that aa^{-1} can be 'cancelled' unless this would result in the symbol's disappearing entirely. For the part aa^{-1} of a symbol containing at least four letters can be eliminated by 'closing up':

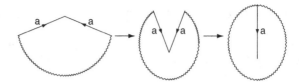

It may look as though this procedure will inevitably introduce curved edges into the diagram. In fact the resulting diagram will still be essentially a graph in the plane, and by Fáry's theorem (1.34) we can redraw it with straight edges and with the same triads of edges spanning triangles as before. In any case the diagram should really be treated as a schema for showing intersections of triangles, and it does not matter whether the edges are straight or curved.

The closing up operation is performed whenever possible during the succeeding stages. As for the symbol aa^{-1} itself, this is one of the final list of standard forms, and is to be regarded as a *sphere* (see diagrams).

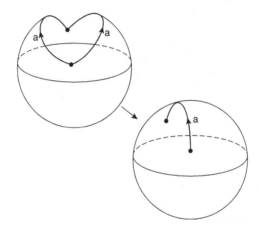

(3) A pair $\ldots a \ldots a \ldots$ is called a *pair of the first kind* and a pair $\ldots a \ldots a^{-1} \ldots$ is called a *pair of the second kind*. A pair of the first kind can be brought together by the following rearrangement:

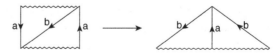

which replaces $\ldots a \ldots a \ldots$ by $\ldots bb \ldots$. This process repeated produces

$$a_1 a_1 a_2 a_2 \ldots a_p a_p X$$

where X consists entirely of pairs of the second kind, or is empty, and we say that $p = 0$ if there are no pairs of the first kind at all. The details of the repetition will be left to the reader, guided by the following hints:

$aab\ldots b\ldots$ can be turned into $aacc\ldots$;
$aab\ldots b^{-1}\ldots$ can be turned into $aab\ldots b\ldots$ provided there is a pair of the first kind besides aa.

The symbol aa represents the closed surface obtained from a circular disk by identifying together each pair of diametrically opposite points on the boundary circle. This cannot be done in $\mathbb{R}^3$ (compare 2.16(5) and 9.18). A surface given by aa is known as a *projective plane* or *sphere with a cross-cap*. Any attempt to draw the surface in $\mathbb{R}^3$ will result in self-intersections, and two such attempts are pictured below. In each case the self-intersection is along the segment XY.

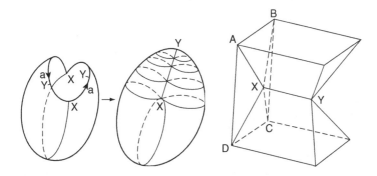

In both models, a is bent double, to lie along XYX. But the two halves, XY and YX, are on different sheets of the surface which intersect accidentally.

The right-hand surface consists of three rectangles (top, bottom and one end), four trapezia and four triangles (ABX and CDX are missing).

(4) If the symbol X in (3) above contains one pair of the second kind then either the whole symbol is aa^{-1} or, in view of (2), there is another pair of the second kind. Choosing $\ldots a\ldots a^{-1}\ldots$ to be the pair which is closest together there must then be two 'interlocking' pairs $\ldots a\ldots b\ldots a^{-1}\ldots b^{-1}\ldots$ (the b^{-1} could also come before a) and this can be turned into $\ldots cdc^{-1}d^{-1}\ldots$ by the following:

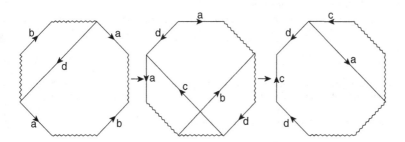

Note that no new pairs of the first kind can appear during this transformation, since no section of the figure, after cutting, is turned over before glueing back. In this way the symbol is turned into[†]

$$a_1a_1 \ldots a_pa_pc_1d_1c_1^{-1}d_1^{-1} \ldots c_qd_qc_q^{-1}d_q^{-1}.$$

The symbol $cdc^{-1}d^{-1}$ represents a *torus* or *sphere with handle* (compare 2.2(3) where c appears as abc and d as fed, reading anticlockwise from bottom left).

(5) Finally, if there is at least one pair of the first kind then every $cdc^{-1}d^{-1}$ (handle) can be turned into two pairs of the first kind (cross caps):

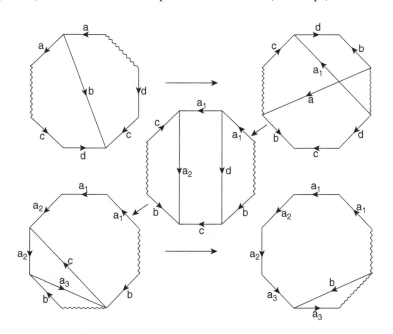

Theorem 2.8. *Every closed surface can (possibly after barycentric subdivisions) be represented by one of the following symbols:*

(1) aa^{-1} *(sphere; orientable),*
(2) $a_1a_1a_2a_2 \ldots a_ka_k$, $k \geq 1$ *(sphere with k cross-caps; non-orientable),*
(3) $a_1b_1a_1^{-1}b_1^{-1}a_2b_2a_2^{-1}b_2^{-1} \ldots a_hb_ha_h^{-1}b_h^{-1}$, $h \geq 1$ *(sphere with h handles; orientable).* □

[†] The symbol given here has the property that the vertices which occur at the ends of the letters in the symbol all represent the same vertex of the closed surface. This comes about because at various points of the reduction pairs aa^{-1} are eliminated (compare (2) above), and a vertex which was on the boundary polygon becomes interior.

The number k or h is called the *genus* of the closed surface; the genus of a sphere is 0.

It is proved in 2.14 that any given closed surface can be reduced to only one of the forms given in 2.8.

When a closed surface is represented by one of the above symbols we say that it is in *standard form*. Two closed surfaces M and N which reduce to the same standard form are called *equivalent*, and we write $M \approx N$. Of course two equivalent closed surfaces may be broken up into triangles in quite different ways. What $M \approx N$ says is that, forgetting about the ways in which M and N are broken into triangles, we can obtain each of them from a plane polygonal region with the same number of sides by making the same identifications, given by the symbols. Thus M and N 'look the same' – are in fact homeomorphic (see Chapter 5).

Any closed surface with standard form aa^{-1} is called a (triangulation of the) *sphere*; likewise closed surfaces with standard forms aa and $aba^{-1}b^{-1}$ are called (triangulations of) *projective plane* and *torus* respectively.

There is a simple geometrical interpretation for the juxtaposition of symbols, such as the succession of torus or 'handle' symbols $aba^{-1}b^{-1}$ which occurs in 2.8(3). Let M and N be closed surfaces given by symbols X and Y respectively (not necessarily in standard form). We shall construct a new surface with symbol XY obtained by juxtaposing X and Y. The new surface is far from being uniquely determined by M and N, but nevertheless its standard form is uniquely determined by those of M and N. This follows from 2.17 below.

2.9. Connected sum of two surfaces

Choose a triangle of M meeting the boundary polygon of the plane representation with symbol X precisely in a vertex of M at the start of X

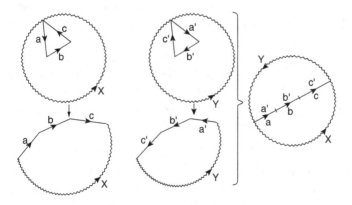

(if necessary barycentrically subdividing M), and similarly with N. Then remove these two triangles and glue together as in the diagram.

The result is a closed surface, orientable if and only if M and N are both orientable. It is called a *connected sum* of M and N. Informally we cut a hole in each surface and glue them together along the boundary rims. Thus for a connected sum of two tori there appear to be two choices:

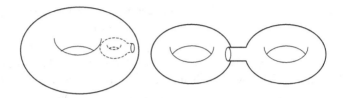

In fact, as already claimed, these have the same standard form (indeed the little torus inside can be pulled through the hole to turn the left-hand picture into the right-hand one!). The surface is called a 'double torus', or 'pretzel'. Thus the surface of 2.8(3) is an 'n-fold torus' which looks like n tori joined by $n-1$ tubes, and 2.8(2) is an 'n-fold projective plane'. In particular the Klein bottle, which has the usual symbol (compare 2.2(4)) $xyx^{-1}y$, has standard form $a_1a_1a_2a_2$, a 'double projective plane'.

Let M and N be closed surfaces. The uniqueness property mentioned before 2.9 (and proved in 2.17) amounts to saying that any two connected sums of M and N are equivalent: we can speak up to equivalence of *the* connected sum of M and N.

Notation 2.10. We write $M \# N$ for any connected sum M and N.

Thus any two choices of $M \# N$ (for given M and N) are equivalent. Indeed, more is true: if M and N are each allowed to vary within an equivalence class then all the possible connected sums are still equivalent. This too is proved in 2.17.

The notation can be extended to several 'summands'; again $M_1 \# \ldots \# M_n$ is well-defined up to equivalence. We write nM for $M \# \ldots \# M$ with n summands. Writing T for any closed surface with standard form $aba^{-1}b^{-1}$, i.e. for any torus, P for any projective plane and S for any sphere, the statement of 2.8 can be rephrased as follows:

every closed surface is equivalent to exactly one of S, kP or hT.

It can be shown that, for any closed surface M, we have $M \# S \approx M \approx S \# M$. This is most easily done using Euler characteristics (described below), but the reader can probably convince himself by drawing pictures. The same goes for the statement $P \# 2P \approx P \# T$ (this is essentially shown in 2.7(5)); note that

$2P \not\approx T$ so that 'cancellation' is not possible. Since $2P$ can be replaced by T provided this leaves at least one P, we can also rephrase 2.8 by saying that

every closed surface is equivalent to exactly one of

$S, hT, P \# \frac{k-1}{2}T$ (k odd), $K \# \frac{k-2}{2}T$ (k even, ≥ 2) *where* $K = 2P$ *is a Klein bottle.*

Euler characteristics

We shall prove that a given closed surface M cannot be reduced to two different standard forms 2.8. In addition to this we shall present a method by which the standard form of any surface given by a polygonal region with pairs of boundary edges identified (compare 2.6) can be found very quickly and without actually performing the scissors and glue operations of 2.7.

Definition 2.11. Let K be a collection of triangles satisfying the intersection condition given before 2.1. Writing $\alpha_0(K), \alpha_1(K), \alpha_2(K)$ for, respectively, the number of vertices, edges and triangles of K, the *Euler characteristic* of K is defined to be

$$\chi(K) = \alpha_0(K) - \alpha_1(K) + \alpha_2(K).$$

Proposition 2.12. *The Euler characteristic of K is unaffected by barycentric subdivision.*

Proof Write the αs after one subdivision as α'; then

$$\alpha'_0 = \alpha_0 + \alpha_1 + \alpha_2, \quad \alpha'_1 = 2\alpha_1 + 6\alpha_2, \quad \alpha'_2 = 6\alpha_2.$$

Hence

$$\alpha'_0 - \alpha'_1 + \alpha'_2 = \alpha_0 - \alpha_1 + \alpha_2. \qquad \square$$

Our immediate aim is to calculate the Euler characteristic of the closed surface corresponding to any given symbol.

Suppose that a closed surface M has a symbol containing n letters (n is necessarily even), and is represented by a plane polygonal region bounded by a simple closed polygon with $n+r$ sides (each side of the polygon corresponding to an edge of M – recall that a letter in a symbol can stand for a succession of edges). The n vertices of the polygon at beginnings or ends of letters will represent some of the vertices of M; let the number of distinct vertices of M amongst these be m.

Then

Theorem 2.13.

$$\chi(M) = m - \frac{1}{2}n + 1.$$

Remark. For the torus in 2.2(3), without any grouping into letters, $n = 12$, $r = 0$, $m = 5$. Grouping *abc* and *fed* into single letters $n = 4$, $r = 8$, $m = 1$ and in both cases $m - \frac{1}{2}n + 1 = 0$. The key fact is that once we have grouped edges into letters each vertex interior to a letter of M is identified only with the corresponding vertex interior to the other appearance of that letter.

Proof of 2.13 The set D of triangles assembled in the plane to give the plane representation of M gives rise to a graph which divides the plane into triangular regions. Write $\alpha_0(D), \alpha_1(D), \alpha_2(D)$ for the numbers of vertices and edges of this graph, and the number of triangular regions, respectively. (D stands for 'disk'.) Euler's formula (1.28) shows that $\alpha_0(D) - \alpha_1(D) + \alpha_2(D) = 1$ since there is one unbounded region. We write $\chi(D)$ for the left-hand side of this (compare 3.33).

The number of vertices of D on the boundary polygon is $n + r$ while the number of vertices of M on the boundary polygon is $m + \frac{1}{2}r$. Since vertices inside the polygon count equally in D and in M, it follows that $\alpha_0(M) - \alpha_0(D) = m - n - \frac{1}{2}r$. Similarly $\alpha_1(M) - \alpha_1(D) = -\frac{1}{2}n - \frac{1}{2}r$ and of course $\alpha_2(M) - \alpha_2(D) = 0$. Thus $\chi(M) - \chi(D) = m - \frac{1}{2}n$ and this, together with $\chi(D) = 1$, gives the result. □

Corollary 2.14. *The Euler characteristic of any closed surface with standard form*

2.8 (1) *is* 2;

2.8 (2) *is* $2 - k$;

2.8 (3) *is* $2 - 2h$.

The standard form of any closed surface M is unique, and is determined by $\chi(M)$ and the orientability or otherwise of M.

Proof For the first statement, it is enough to calculate the Euler characteristics of the closed surfaces given by 2.8, since (as the reader may verify) the scissors and glue operations of 2.7, as well as barycentric subdivision, do not affect the Euler characteristic. In case (1), $m = 2$, while in cases (2) and (3), $m = 1$. The result now follows from the theorem.

For the second statement, note that the scissors and glue of 2.7 does not affect orientability either. Since no two standard forms agree in point of orientability and Euler characteristic, the result follows. □

Remark 2.15. When calculating the number m of 2.13 the procedure of the following example is used. Choose any vertex (at beginning or end of a letter) and call it A, say $A_1 = A$.

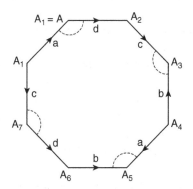

A_1 is at the end of a; so is A_5. Hence in M, $A_5 = A$.
A_5 is at the end of b; so is A_3. Hence in M, $A_3 = A$.
A_3 is at the end of c; so is A_7. Hence in M, $A_7 = A$.
A_7 is at the beginning of d; so is A_1: we are back where we started.

The important thing is that no other vertex can be A, since in M the link of A is one simple closed polygon, which is indicated by the dotted lines in the diagram. How start at another vertex, say $A_2 = B$ and deduce $A_4 = A_6 = A_8 = B$. Hence $m = 2$, so that $\chi(M) = -1$ and since M is non-orientable (it contains $\ldots b \ldots b \ldots$), M has standard form $3P$, i.e. $a_1 a_1 a_2 a_2 a_3 a_3$. Thus M is a sphere with three cross-caps.

Examples 2.16. Check the following, (1)–(4).

(1X) $abcbca$ is a Klein bottle, i.e. has standard form $2P$.
(2X) $abca^{-1}cb^{-1}$ has standard form $3P$.
(3X) $abcdefe^{-1}db^{-1}afc$ has standard form $6P$.
(4X) $ae^{-1}a^{-1}bdb^{-1}ced^{-1}c^{-1}$ has standard form $2T$.

★(5X) Since any orientable closed surface is equivalent to a connected sum of tori (see the discussion following 2.10) it is clear that a representative of each equivalence class can be realized in $\mathbb{R}^3$. In Chapter 9 (see 9.18) we shall sketch a proof of the theorem that no non-orientable closed surface can be realized in $\mathbb{R}^3$. However they can all be realized in $\mathbb{R}^4$, as follows. Since every non-orientable closed surface is equivalent to a connected sum of tori and either a

58   *Closed surfaces*

protective plane P or a Klein bottle K it is enough to check that P and K can be realized in $\mathbb{R}^4$,

For P, start with a Möbius band, triangulated as in the diagram.

It is clear that with enough triangles this can be realized in $\mathbb{R}^3$ (can the reader discover the minimum number needed?). Now take coordinates (x, y, z, t) in $\mathbb{R}^4$ and let the Möbius band be realized in the space $t = 0$. We add triangles as follows: for each pair v, w of vertices consecutive along the rim of the band, add the triangle with vertices v, w and $(0, 0, 0, 1)$. The result is a closed surface (for example the link of $(0, 0, 0, 1)$ is the rim of the band, which is a simple closed polygon) and the Euler characteristic is easily verified to be 1. Hence it is a projective plane. We have 'added a cone' to the boundary rim of the band. Cones will be used on several occasions later in the book; see the index for references.

The Klein bottle K can be constructed in $\mathbb{R}^4$ in a similar way, starting with a triangulation of the 'double Möbius band' – two bands with parts of their rims identified. The boundary rim of the double band is still a single closed curve, and adding a cone gives a non-orientable closed surface of Euler characteristic 0, hence a Klein bottle.

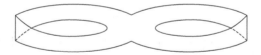

An alternative approach is to start with the picture of a Klein bottle in 2.2(4), where there is self-intersection along a circle. Using the fourth spatial dimension one of the-intersecting tubes can be 'moved aside' to miss the other. (This is analogous to taking two intersecting lines in a horizontal plane and moving one of them slightly upwards to miss the other.) ★

★(6X) Let M be a closed surface in $\mathbb{R}^3$, not passing through the origin 0, and with the property that every half-line (ray) through 0 meets M precisely once. Then M is a triangulation of the sphere, i.e. has standard form aa^{-1}. One way to see this is as follows. For each vertex v of M let v_1 be the (unique) intersection of the half-line $0v$ with a fixed sphere centred at 0. Thus $v \to v_1$ projects the vertices of M on to the sphere. Similarly the edges of M project to arcs of great circles on the sphere and triangles to spherical triangles. Since half-lines meet M exactly once we obtain a collection of (spherical) triangles covering the sphere, with intersections exactly as in M. To recover the standard form aa^{-1} we can remove one spherical triangle t and flatten separately t, and the rest of the sphere, on to a plane (e.g. by stereographic projection). This gives a plane representation of M by t and a triangular region which is itself broken up into triangles. Re-identifying one edge of t with one boundary edge of the triangular region gives the standard form aa^{-1}. It is possible to avoid curved triangles in this argument, though of course in the plane they can in any case be straightened by Fáry's theorem (1.34). ★

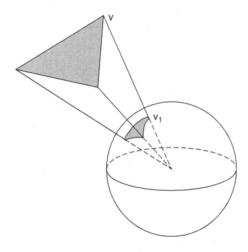

★(7X) Suppose that the closed surface M_1 in (6) has the property that it is 'symmetric about 0', i.e. that it is left unchanged by reflexion through the point 0. Then, at any rate provided M_1 has enough triangles, we can form a closed surface M_2 by identifying opposite triangles of M_1. Since $\chi(M_1) = 2$ it is clear that $\chi(M_2) = 1$, so that M_2 must be a triangulation of the projective plane. To see this informally, think of identifying diametrically opposite points all over the sphere. We can throw away the southern hemisphere provided we leave the equator and identify opposite points of it. Flattening out the hemisphere we get a disk, or a square, with opposite points of its boundary identified. The latter is

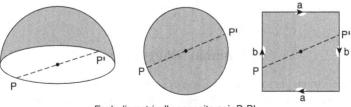

Each diametrically opposite pair P, P' is identified to a point

one of the usual pictures of a projective plane (2.2(4)). ★

★(8XX) Let M be a closed surface. We form a *subdivision* M_1 of M, by dividing up every triangle of M into smaller triangles. We require the subdivisions of any two adjacent triangles to be consistent on their common edge as in the diagram. A special subdivision is the barycentric subdivision used in this chapter. (See also Chapter 5.) Let $\overline{\alpha}_0(M_1), \overline{\alpha}_1(M_1)$ denote the numbers of vertices and edges of M_1 lying in edges of M. Show that $\overline{\alpha}_0(M_1) - \overline{\alpha}_1(M_1) = \alpha_0(M) - \alpha_1(M)$. Note that, if D denotes the part of M_1 lying within a single triangle of M, then $\chi(D) = 1$; compare the proof of 2.13. Deduce that $\chi(M) = \chi(M_1)$: this says that the Euler characteristic is invariant under subdivision. ★

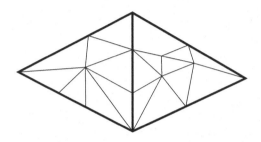

Now consider the effect of making a connected sum (2.9) on Euler characteristics. Since barycentric subdivision does not affect Euler characteristics (2.12) it is immediate from the definition that $\chi(M_1 \# M_2) = \chi(M_1) + \chi(M_2) - 2$ for any connected sum $M_1 \# M_2$ of two closed surfaces M_1 and M_2. It is also true that $M_1 \# M_2$ is orientable if and only if M_1 and M_2 are both orientable: thus orientability and Euler characteristic for $M_1 \# M_2$ are determined by M_1 and M_2. By 2.14 we have:

Proposition 2.17.

(1) *Any two connected sums of M_1 and M_2 are equivalent.*
(2) *If $M_1 \approx N_1$, and $M_2 \approx N_2$ then*
$$M_1 \# M_2 \approx N_1 \# N_2.$$
□

★ Minimal triangulations

Among all triangulations of, say, a torus – that is, among all closed surfaces M with standard form $aba^{-1}b^{-1}$ – there will be some which have the fewest possible triangles. In this section we investigate such 'minimal triangulations' briefly and relate them to a famous problem of graph theory.

Let us fix attention on a particular standard form, which has Euler characteristic χ given by 2.14. Any closed surface M with this standard form has the following properties:

2.18
$$\begin{cases} \alpha_1(M) \leq \tfrac{1}{2}\alpha_0(M)(\alpha_0(M)-1) & \text{(see the note following 1.1)} \\ 3\alpha_2(M) = 2\alpha_1(M) & \text{(2.2 (2))} \\ \alpha_0(M) - \alpha_1(M) + \alpha_2(M) = \chi & \text{(2.11)} \end{cases}$$

These give $\alpha_0^2 - 7\alpha_0 + 6\chi \geq 0$. With the exceptions noted below, this is equivalent to

2.19
$$\alpha_0 \geq \tfrac{1}{2}\left(7 + \sqrt{49 - 24\chi}\right).$$

The exceptions are as follows:

If $\chi = 1$ there is a solution $(\alpha_0, \alpha_1, \alpha_2) = (1, 0, 0)$
If $\chi = 2$ there are solutions $(\alpha_0, \alpha_1, \alpha_2) = (2, 0, 0)$ and $(3, 3, 2)$.

However none of these exceptions corresponds to a genuine surface. The only one that makes sense at all is $(3, 3, 2)$ for the sphere, and this is two distinct triangles with the same three vertices and the same three edges, a configuration which cannot be realized with flat triangles and which violates the intersection condition given before 2.1.

Note that $\alpha_0(M)$ is minimal if and only if $\alpha_2(M)$ is minimal, and also that equality occurs in 2.19 if and only if it occurs in the first relation of 2.18.

Examples 2.20. (1) Consider the standard form aa^{-1}: the sphere. Then $\chi = 2$ and from 2.19, $\alpha_0(M) \geq 4$. In fact there is a triangulation, namely the hollow tetrahedron, with exactly four vertices.

(2) Consider the standard form aa: the projective plane. Then $\chi = 1$ and $\alpha_0(M) \geq 6$. A triangulation with six vertices is drawn in 4.17(5).

(3) Consider the standard form $aba^{-1}b^{-1}$: the torus. Then $\chi = 0$ and $\alpha_0(M) \geq 7$. A triangulation with seven vertices is drawn in 3.22(1). There is another standard form with $\chi = 0$, namely $a_1a_1a_2a_2$: the Klein bottle. Is there a triangulation with seven vertices? In fact not: the smallest possible number is eight, and it is not hard to find triangulations with eight vertices. This is a curiously exceptional example, for in every other case where the right-hand side of 2.19 is an integer there does exist a triangulation with exactly that number of vertices. Not surprisingly, this is far from easy to prove; we shall shortly sketch a proof which depends on a very difficult theorem in graph theory, and also on a result to be proved in Chapter 9.

(4X) Show from 2.19 that

$$\alpha_0(M) \geq \left\lfloor \tfrac{1}{2}\left(9 + \sqrt{48 - 24\chi}\right) \right\rfloor$$

where $\lfloor x \rfloor$ means the greatest integer $\leq x$.

(5) In the non-orientable case, Ringel (1955) has determined precisely the number of vertices in a minimal triangulation. It is $\left\lfloor \tfrac{1}{2}\left(9 + \sqrt{48 - 24\chi}\right) \right\rfloor$ for $\chi \neq 0, -1$; and 8 for $\chi = 0$ (Klein Bottle); 9 for $\chi = -1$. For the orientable case Jungerman & Ringel (1980) establish a formula for the number of vertices in a minimal triangulation, namely $\left\lceil \tfrac{1}{2}(7 + \sqrt{49 - 24\chi}) \right\rceil$ where $\lceil \; \rceil$ is the 'ceiling' function taking a real number x to the smallest integer $\geq x$. There is an exception here too: for $\chi = -2$ the number is 10.

Given a graph G we can enquire, for example, 'Is it possible to embed G in the torus?' This means 'Is there some triangulation M of a torus with the property that the graph consisting of the set of vertices and edges of M contains a subgraph isomorphic to G?' Thus for example let G_n denote the complete graph with n vertices (every pair of vertices being joined by an edge); then for $n \leq 7$, G_n can be embedded in a torus, for the triangulation drawn in 3.22(1) has the property that its set of vertices and edges is a G_7. Indeed, whenever we have a triangulation for which equality holds in 2.19 it will satisfy $\alpha_1 = \tfrac{1}{2}\alpha_0(\alpha_0 - 1)$, and therefore the set of vertices and edges of the triangulation will be a complete graph G_{α_0}.

There is an alternative, less formal, way in which we can think of embedding a graph in a closed surface. We can think of the surface in its usual rounded shape and the graph drawn with curved edges in the surface. For example the left-hand diagram below shows G_5 in the torus. In fact, the two ideas are equivalent: a graph can be embedded in this way in a curved surface if and only if it can be embedded in the way previously defined as a straight graph contained in some triangulation. Roughly speaking, given a graph in a curved surface, we

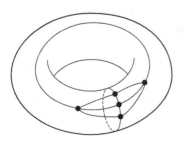 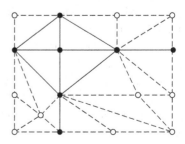

add extra vertices and edges away from the graph until the surface is divided up into curved triangles, and then regard this as a schema for the construction of a genuine (straight) triangulation (see 2.6). This is analogous to the construction of a triangular graph in the plane (1.34(I)) except for the introduction of extra vertices. One way of carrying this out for G_5 in the torus is illustrated above.

In the spirit of the preceding paragraph, let us see how to embed an arbitrary graph in a closed surface of sufficiently high genus (2.8). Given a graph G in $\mathbb{R}^3$ (see 1.4), project G from some point in $\mathbb{R}^3$ on to a plane, choosing the point of projection O in such a way that the ray through O and a vertex of G does not meet G again, while the ray through O and any other point of G meets G again at most once. For example, starting with G_5 we might obtain the diagram in 1.5(1) in the plane. At each crossing point (not the projection of a vertex) we stick a small handle on the plane:

In this way all the crossings are eliminated, and enclosing the projected graph in a large circle which is then capped by a hemisphere we obtain a graph isomorphic to G realized in a sphere with handles (or with crosscaps, if a crosscap is added somewhere). Applying this procedure to 1.5(1) we can embed G_5 in an orientable surface of genus 5 – far from the best result, for it also embeds in the torus, which has genus 1.

We are interested, in what follows, not in whether a graph embeds in a particular surface, but in whether it embeds in some surface in a given equivalence class – the notion of equivalence being that given after 2.8 (i.e. having the same standard form). Thus unless otherwise stated we regard all surfaces as variable within a definite equivalence class. This most accurately reflects the informal idea of drawing the graph on a curved surface.

It is not an easy matter to determine the surface of smallest genus in which a given graph can be embedded. A remarkable theorem, due essentially to G. Ringel and J. W. T. Youngs, but in which several other people had a hand, determines just this for the complete graph G_n. (See Ringel & Youngs (1968), Ringel (2002) and the books of Ringel, Bollobás and Gross & Yellen listed in the References.) The following statement of the theorem is directed at the opposite problem: given a surface (variable of course within an equivalence class) what is the largest complete graph which can be embedded in it?

Theorem 2.21 (Ringel–Youngs). *The largest complete graph which will embed in a closed surface, other than the Klein bottle, of Euler characteristic χ, is the one with*

$$\mathcal{H}(\chi) = \left\lfloor \tfrac{1}{2}\left(7 + \sqrt{49 - 24\chi}\right) \right\rfloor$$

vertices (as usual $\lfloor \ \rfloor$ stands for the integer part). In the case of the Klein bottle, the largest complete graph is G_6, not G_7. □

Note the anomalous position of the Klein bottle, and also the similarity between 2.21 and 2.19. I claim, in fact, the following.

Theorem 2.22. [†] *Suppose that $\tfrac{1}{2}(7 + \sqrt{49 - 24\chi})$ is an integer, λ say, where χ is an integer $\leqslant 2$. Then there is a closed surface with λ vertices and Euler characteristic χ. Indeed if $\chi < 0$ is even then there are two such, one orientable and one non-orientable. If $\chi = 0$ then there is only one, which is a torus. The set of vertices and edges of the surface is a complete graph G_λ.*

Note that here the claim is that G_λ embeds in a particular closed surface, not merely one of a given equivalence class. Note also that the triangulation is minimal by 2.19. In fact the theorem above asserts that the lower bound of 2.19 is attained provided $\tfrac{1}{2}(7 + \sqrt{49 - 24\chi})$ is an integer.

Here is a sketch of a proof of 2.22. It follows from the Ringel–Youngs theorem that, with $\chi \neq 0$, G_λ embeds in any closed surface with characteristic χ but not in any with characteristic $> \chi$. The case $\chi = 0$ requires special attention. From this it can be shown that whenever G_λ is embedded in a closed surface of characteristic χ, the graph will divide it up into triangles: that is, G_λ forms the vertices and edges of triangulation in which the triangles are the triangular regions into which G_λ divides the surface. A proof of this is sketched in the last paragraph but one of 9.28(7), which depends on several results of Chapter 9. Using the fact that G_λ is a graph there is no difficulty in checking that the triangles satisfy the intersection condition given before 2.1. (The only possible exception is G_3, embedded in the sphere to give two (curved) triangles meeting in all three edges, but fortunately $\lambda = 3$ is not a possible value. This anomalous case turned up in the discussion of 2.19 as well.) □

Examples 2.23. (1) Take $\chi = -10$ in 2.22. Then $\lambda = 12$; hence there are surfaces equivalent to $6T$ and to $12P$ (both of which have characteristic -10)

[†] My aim in this brief exposition is to display some connexions between the Ringel–Youngs theorem, minimal triangulations and (shortly) colouring problems. I hasten to point out that existing proofs of the Ringel–Youngs theorem proceed on a case by case basis, and that one case is close in content to the present theorem.

having 12 vertices. In each case the set of vertices and edges is G_{12}: this result says that starting with G_{12} we can build up both an orientable and a non-orientable closed surface by filling in suitable triangles. (The G_{12} had better be in a high-dimensional space to avoid accidental intersections of triangles. In fact a theorem in Chapter 3 (3.19) says that five dimensions will do.)

(2X) For which values of χ is $\frac{1}{2}(7 + \sqrt{(49 - 24\chi)})$ an integer? (Answer: $\chi = (7n - n^2)/6$ where $n \equiv 0, 1, 3$ or $4 \pmod 6$.)

(3X) Use 2.21 to find explicitly the Euler characteristics of the orientable and non-orientable surfaces of lowest genus into which it is possible to embed G_n.

It is worth pointing out the connexion between 2.21 and the *colouring problem* for graphs in a closed surface. A graph is n-*colourable* if each vertex can be assigned one of n colours in such a way that two vertices at the ends of an edge are always coloured differently. Thus for example a graph with n vertices is certainly n-colourable, and a complete graph G_n is not m-colourable for any $m < n$. Now consider all graphs which can be embedded in a given closed surface (or, as usual, in any equivalent one). What is the smallest n for which all these graphs are n-colourable? This number is called the *chromatic number* of the closed surface. Clearly, from 2.21 the chromatic number of any closed surface of Euler characteristic χ is at least $\mathcal{H}(\chi)$ (except for the Klein bottle, when it is at least 6.) In fact Heawood (1890) proved that the chromatic number is $\leqslant \mathcal{H}(\chi)$, for any closed surface except the sphere. (See 9.28(8).) Heawood actually believed, incorrectly, that he had proved the chromatic number equal to $\mathcal{H}(\chi)$, and the equality became known as *Heawood's conjecture*. Franklin (1934) found a counterexample, when he showed that the chromatic number of the Klein bottle is 6, not $\mathcal{H}(0) = 7$. The Ringel–Youngs theorem showed, astonishingly, that Franklin's counterexample was the only one. Thus the chromatic number is

$\mathcal{H}(\chi)$ for any closed surface except Klein bottle or sphere

6   for the Klein bottle

$\geqslant 4$ for the sphere.

The sphere is not covered by Heawood's theorem: in fact in his 1890 paper Heawood showed only that the chromatic number of a sphere is $\leqslant 5$. It had been conjectured around 1852 that the chromatic number was 4 and indeed another function of Heawood's paper was to expose a fallacious proof of this conjecture. For over 100 years the notorious 'four colour conjecture' remained unsolved. A proof was found in June 1976 and published the following year, but even then the proof used high speed computers and it is a matter for speculation whether a purely conceptual proof that can be checked by a human being will ever be found. At any rate the *four colour theorem* asserts that the chromatic number of the sphere is 4. This is the same as for graphs in the plane.

For an account of the proof by the original authors see Appel & Haken (1977); see also the book *Four Colours Suffice* by Wilson listed in the References. For the five colour theorem see for example the book *Introduction to Graph Theory* by Wilson. For the history of the four colour conjecture, and the connexion with geographical maps, see the book of Biggs, Lloyd & Wilson.

3
Simplicial complexes

The reader may have noticed that although surfaces are constructed out of triangles I did not define the word 'triangle' in Chapter 2. Doubtless the reader has come across triangles before. Nevertheless a precise definition in terms suited to our purpose is given below, where triangles appear under the alias of '2-simplexes'. The precise definition makes it clear that 'triangle' is a good way to continue the sequence 'point, segment, ...' (which becomes '0-simplex, 1-simplex, ...') and suggests that the fourth term should be 'solid tetrahedron': this is our three-dimensional building block. We shall need building blocks in higher dimensions too, but most examples in the text will be, as hitherto, two-dimensional. Algebraic concepts closely analogous to the cycles of Chapter 1 will be introduced in Chapter 4.

It is possible, by judicious skipping in this chapter, to avoid contact with simplexes of dimension higher than two. All initial definitions are stated for the case of two dimensions as well as in general; from 3.9 just assume $n \leqslant 2$.

Everything in the chapter takes place in a real vector space $\mathbb{R}^N$. Elements of $\mathbb{R}^N$ are called *points* or *vectors*. Unless otherwise stated the only restriction on N is that it should be large enough for the discussion to make sense; thus if we speak of four non-coplanar points then obviously N must be at least 3.

Simplexes

It is clear that two points v^0 and v^1 are the end-points of a segment if and only if they are distinct. The segment then consists of all points $(1 - \lambda)v^0 + \lambda v^1$ where $0 \leqslant \lambda \leqslant 1$, i.e. it is the set of points

$$\{\lambda_0 v^0 + \lambda_1 v^1 : \lambda_0 + \lambda_1 = 1, \quad \lambda_0 \geqslant 0, \quad \lambda_1 \geqslant 0\}.$$

Similarly three points v^0, v^1, v^2 are the vertices of a triangle if and only if they are not collinear, i.e. if and only if $v^1 - v^0$ and $v^2 - v^0$ are linearly independent. In the left-hand diagram

$$P = \lambda v^0 + (1 - \lambda)v^1 \quad (0 \leqslant \lambda \leqslant 1),$$

$$Q = \mu(\lambda v^0 + (1 - \lambda)v^1) + (1 - \mu)v^2 \quad (0 \leqslant \mu \leqslant 1).$$

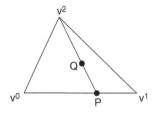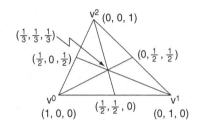

Hence

$$Q = \lambda_0 v^0 + \lambda_1 v^1 + \lambda_2 v^2$$

where $\lambda_0 = \lambda\mu, \lambda_1 = (1 - \lambda)\mu$ and $\lambda_2 = 1 - \mu$. Notice that the λ_i are $\geqslant 0$ and have sum 1. Conversely given non-negative $\lambda_0, \lambda_1, \lambda_2$ with sum 1 we can solve for λ and μ; hence the triangle is the set of points

$$\{\lambda_0 v^0 + \lambda_1 v^1 + \lambda_2 v^2 : \lambda_0 + \lambda_1 + \lambda_2 = 1, \ \lambda_i \geqslant 0 \ \text{ for } i = 1, 2, 3\}.$$

The advantage of the λ_i over λ and μ is that any given point Q has uniquely determined λ_i (compare 3.5) whereas v^2 has $\mu = 0$ but arbitrary λ. The numbers $\lambda_0, \lambda_1, \lambda_2$ are called the *barycentric coordinates* of Q. The barycentric coordinates of some points in the triangle are given in the right-hand diagram above.

It can be shown that if we drop the condition of non-negativeness on the λ_i the above set of points is enlarged to include all those in the plane of the triangle. Points with all $\lambda_i > 0$ are inside the triangle, points with two $\lambda_i > 0$ and one zero are on an edge, and points with two λ_i equal to zero are at a vertex.

Definition 3.1. Let $v^0, \ldots, v^n$ be $n + 1$ vectors in $\mathbb{R}^N$, $n \geqslant 1$. They are called *affinely independent* (*a-independent*) if $v^1 - v^0, \ldots, v^n - v^0$ are linearly independent. By convention if $n = 0$ then the vector v^0 is always *a*-independent (even if it is zero).

Simplexes

Examples 3.2. (1) v^0 and v^1 are a-independent if and only if $v^0 \neq v^1$; v^0, v^1, v^2 are a-independent if and only if they are not collinear.

(2X) Let $v^0, \ldots, v^n$ be in $\mathbb{R}^N$. They are a-independent if and only if, for all real numbers $\lambda_0, \ldots, \lambda_n$, we have

$$\begin{pmatrix} \lambda_0 v^0 + \cdots + \lambda_n v^n = 0 \\ \lambda_0 + \cdots + \lambda_n = 0 \end{pmatrix} \Rightarrow \lambda_0 = \cdots = \lambda_n = 0.$$

[Hint: $\lambda_0 v^0 + \cdots + \lambda_n v^n = 0 = \lambda_1(v^1 - v^0) + \cdots + \lambda_n(v^n - v^0)$ when $\Sigma \lambda_i = 0$.] This makes it clear that the definition of a-independence does not in reality depend on the ordering of the v^i: we could equally well define it by the linear independence of $v^0 - v^1, v^2 - v^1, v^3 - v^1, \ldots, v^n - v^1$.

Also, adding a constant to all the v^i does not affect their a-independence or otherwise. (In contrast, even if v and w are linearly independent, $v - v$ and $w - v$ are not.) Thus a-independence does not depend on the choice of origin.

★(3) Let U be a subspace of $\mathbb{R}^N$ and let $b \in \mathbb{R}^N$, the set of vectors

$$U + b = \{u + b : u \in U\}$$

is called an *affine subspace* of $\mathbb{R}^N$ of *dimension* equal to that of U. We say that $U + b$ is *parallel* to U (it certainly has no points in common with U provided $b \notin U$, and if $b \in U$ then $U + b = U$).

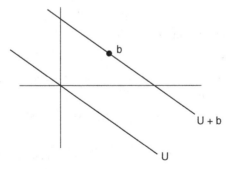

For example, let $N = 2$ and let U be a line through the origin. Then $U + b$ is the line parallel to U, through b.

Notice that, given $U + b$, we can recover U:

$$U = \{v - w : v, w \in U + b\}.$$

On the other hand b is not uniquely determined by $U + b$ since in fact

$$U + b = U + b' \Leftrightarrow b - b' \in U.$$

(For the example above, these say that we can recover U by moving $U + b$ parallel to itself until it passes through the origin; and that replacing b by any $b' \in U + b$ gives the same line parallel to U.)

If $v^0, \ldots, v^n$ are a-independent then $v^1 - v^0, \ldots, v^n - v^0$, being linearly independent, are a basis for a subspace U of $\mathbb{R}^N$ of dimension n. The affine subspace $U + v^0$ then has dimension n and contains all of $v^0, \ldots, v^n$. Notice that any affine subspace $U' + b$ containing $v^0, \ldots, v^n$ must have dimension $\geq n$, for U' will contain the n linearly independent vectors $v^1 - v^0, \ldots, v^n - v^0$. Hence the smallest affine subspace containing $v^0, \ldots, v^n$ has dimension n.

Definition 3.1 can be reformulated: $v^0, \ldots, v^n$ are a-independent if and only if the smallest affine subspace of $\mathbb{R}^N$ containing them has dimension n.

★

Definition 3.3. Let $v^0, \ldots, v^n$ be vectors in $\mathbb{R}^N$. A vector v is said to be *affinely dependent (a-dependent)* on them if there exist real numbers $\lambda_0, \ldots, \lambda_n$ such that

$$\lambda_0 + \cdots + \lambda_n = 1 \quad \text{and} \quad v = \lambda_0 v^0 + \cdots + \lambda_n v^n.$$

Examples 3.4. (1X) If $v^0, \ldots, v^n$ are a-independent then none of them is a-dependent on the rest. If v is a-dependent on $v^0, \ldots, v^n$ then $v^0, \ldots, v^n, v$ are not a-independent.

★(2) The set of vectors a-dependent on $v^0, \ldots, v^n$ is the smallest affine subspace of $\mathbb{R}^N$ containing $v^0, \ldots, v^n$ and is called the *affine span* of $v^0, \ldots, v^n$. ★

Proposition 3.5. *Let $v^0, \ldots, v^n$ be a-independent and let v be a-dependent on them. Then there exist unique real numbers $\lambda_0, \ldots, \lambda_n$ such that $\Sigma \lambda_i = 1$ and $v = \Sigma \lambda_i v^i$.* The λ's are called the barycentric coordinates *of v with respect to $v^0, \ldots, v^n$.*

Proof The λ's exist by 3.3. To prove uniqueness suppose that $\Sigma \lambda_i v^i = \Sigma \mu_i v^i$ where $\Sigma \lambda_i = \Sigma \mu_i = 1$ (all sums run from 0 to n). Then $\Sigma \nu_i v^i = 0$, where $\nu_i = \lambda_i - \mu_i$ so that $\Sigma \nu_i = 0$. Hence by 3.2(2) we have $\nu^0 = \cdots = \nu^n = 0$, i.e. $\lambda_i = \mu_i$ for $i = 0, \ldots, n$. □

By analogy with the discussion preceding 3.1 we can now define the general concept of simplex.

Definition 3.6. Let $v^0, \ldots, v^n$ be a-independent. The (closed) *simplex* with vertices $v^0, \ldots, v^n$ is the set of points a-dependent on $v^0, \ldots, v^n$ and with every barycentric coordinate ≥ 0. The simplex is also said to be *spanned*

Simplexes   71

by $v^0, \ldots, v^n$. The points with all barycentric coordinates > 0 are said to be *interior* to the simplex, and the set of interior points is sometimes called the *open simplex* with vertices $v^0, \ldots, v^n$. The *boundary* of the simplex consists of those points which are not interior, i.e. which have at least one barycentric coordinate $= 0$.

Notation 3.7. In this book, the unqualified word 'simplex' will mean 'closed simplex'. The simplex with a-independent vertices $v^0, \ldots, v^n$ will be denoted by $(v^0 \ldots v^n)$ and often represented by a symbol like s_n or just s. The integer n is called the *dimension* of $(v^0 \ldots v^n)$ (compare 3.2(3)). A simplex of dimension n is called an *n-simplex*. The boundary of s_n will be denoted by $\dot{s}_n$ and the interior by $\overset{\circ}{s}_n$.

Examples 3.8. (1) For any vector v^0 the 0-simplex (v^0) is just the set $\{v^0\}$. We often denote (v^0) by merely v^0. The open 0-simplex with vertex v^0 is also $\{v^0\}$. The boundary of (v^0) is empty.

For any two distinct vectors v^0, v^1 the 1-simplex $(v^0 v^1)$ is the closed *segment* with end-points v^0, v^1. The boundary of $(v^0 v^1)$ is $\{v^0, v^1\}$, and the interior is the open segment with end-points v^0, v^1.

For any three non-collinear vectors v^0, v^1, v^2 the 2-simplex $(v^0 v^1 v^2)$ is the *triangle* with vertices v^0, v^1, v^2. The boundary of $(v^0 v^1 v^2)$ is the set of points on the edges of the triangle, i.e. $(v^0 v^1) \cup (v^1 v^2) \cup (v^2 v^0)$, and the interior is the rest of $(v^0 v^1 v^2)$.

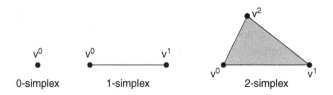

0-simplex   1-simplex   2-simplex

For any four non-coplanar vectors v^0, v^1, v^2, v^3 the 3-simplex $(v^0 v^1 v^2 v^3)$ is the *solid tetrahedron* with vertices v^0, v^1, v^2, v^3. Its boundary consists of four triangles.

★(2) A closed simplex is a closed subset of $\mathbb{R}^N$ in the usual sense that it contains all its limit points. This can be seen by means of barycentric coordinates: the ith barycentric coordinate of each point of a sequence of points in the simplex will be a real number $\geqslant 0$, so that the ith barycentric coordinate of the limit point of a convergent sequence will also be $\geqslant 0$. The interior of a closed simplex is the corresponding open simplex, which is not an open subset of $\mathbb{R}^N$ unless $n = N$. ★

(3X) A simplex determines its vertices – that is, if $(v^0 \ldots v^n) = (w^0 \ldots w^m)$ then $m = n$ and the v's are merely a permutation of the w's. To prove this we must find a characterization of the vertices of a simplex purely in terms of the set of points in the simplex. (This is perhaps a slightly unfamiliar situation – in euclidean geometry a triangle is defined in terms of its sides, and the sides intersect in pairs at the vertices, which are therefore determined. Here, on the other hand, a triangle is defined by the set of points inside it and on its boundary.) Such a characterization, which can be established using barycentric coordinates, is this: a point $v \in (v^0 \ldots v^n)$ is a vertex (i.e. is one of the points $v^0, \ldots, v^n$) if and only if it is not the mid-point of any segment joining two distinct points of $(v^0 \ldots v^n)$.

(4X) A simplex is *convex*, that is if v and w are points of the simplex then every point of the (straight) segment joining v and w belongs to the simplex. Indeed the simplex is the smallest convex set containing its vertices, which is expressed by saying that it is the *convex hull* of its vertices.

(5X) Let $w \in (v^0 \ldots v^n)$ and let $v \neq w$ be an interior point of $(v^0 \ldots v^n)$. Show that the segment from v to w, produced beyond w if necessary, meets the boundary of the simplex in precisely one point. (This amounts to showing that there is precisely one point of the simplex of the form $(1 - \alpha)v + \alpha w$ with $\alpha \geqslant 1$ and having at least one barycentric coordinate zero.)

(6X) Let v be an interior point of the simplex $(v^0 \ldots v^n)$. Show that $v^0, \ldots, v^{n-1}, v$ are *a*-independent.

Definition 3.9. Let $s_n = (v^0 \ldots v^n)$ be a simplex. A *face* of s_n is a simplex whose vertices form a (non-empty) subset of $\{v^0, \ldots, v^n\}$. If the subset is proper (i.e. not the whole of $\{v^0, \ldots, v^n\}$) we say that the face is a *proper face*. If s_p is a face of s_n we write $s_p < s_n$ or $s_n > s_p$; note that $s_n < s_n$ for any simplex s_n.

Remarks and Examples 3.10. (1) The boundary of s_n (3.6) is the union of the proper faces of s_n.

(2) A face of s_n consists of those points of s_n for which the barycentric coordinates attached to vertices not in the face are zero. It follows at once that two faces of s_n either are disjoint or else intersect in a face of s_n.

(3) Strictly speaking a vertex v^i of s_n is not a face of s_n. What is a face is the 0-simplex $(v^i) = \{v^i\}$. It does little harm, however, to fail to distinguish between v^i and (v^i).

(4X) An n-simplex has $\binom{n+1}{p+1}$ faces of dimension p and $2^{n+1} - 1$ faces altogether.

(5) The zero- and one-dimensional faces of an n-simplex form a complete graph on $n + 1$ vertices.

Ordered simplexes and oriented simplexes

A fairly detailed treatment of orientation is given in an appendix to this chapter, which is intended to motivate Definition 3.11 below and relate it to the concept of orientation in vector spaces. It is not necessary to read the appendix in order to follow this section. Orientations of 2-simplexes have already been introduced informally in connexion with closed surfaces (see 2.3). The algebraic theory developed from Chapter 4 onwards depends strongly on the concept of orientation, which is used to define a 'boundary' homomorphism analogous to that introduced in 1.19 for oriented graphs. (Compare 1.21(3).)

According to Definition 3.6, permutation of the vertices of a simplex leaves the simplex unchanged. If on the other hand we insist on a definite ordering of the vertices the result is called an *ordered simplex*. Thus associated with any n-simplex there are $(n + 1)!$ distinct ordered simplexes. In practice we do not wish to distinguish between all of these.

Consider for example a 2-simplex $(v^0 v^1 v^2)$. The six ordered simplexes associated with it are those in which the ordering is as follows:

$$(v^0 v^1 v^2) \quad (v^0 v^2 v^1)$$
$$(v^1 v^2 v^0) \quad (v^2 v^1 v^0)$$
$$(v^2 v^0 v^1) \quad (v^1 v^0 v^2)$$

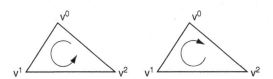

The three even permutations of (v^0, v^1, v^2) all order the vertices 'anti-clockwise' (taking this as the direction for (v^0, v^1, v^2) itself) while the three odd permutations order them 'clockwise'. This suggests the following general definition.

Definition 3.11. Let $s_n = (v^0 \ldots v^n)$ be an n-simplex. An *orientation* for s_n is a collection of orderings for the vertices consisting of a particular ordering and all even permutations of it. An *oriented n-simplex* σ_n is an n-simplex s_n together with an orientation for s_n.

Notation 3.12. We write $\sigma_n = (v^0 \ldots v^n)$ to mean that σ_n is the oriented n-simplex with vertices $(v^0, \ldots, v^n)$ and orientation given by the ordering displayed and all even permutations of it. It will always be clear from the context whether $(v^0 \ldots v^n)$ means a simplex or an oriented simplex. When $n > 0$ we write $-\sigma_n$ for the *oriented simplex* consisting of the same n-simplex and the collection of other possible orderings of the vertices. Thus by definition $-(-\sigma_n) = \sigma_n$.

Remarks 3.13. (1) A 0-simplex has a unique orientation; an n-simplex for $n \geqslant 1$ has two orientations.

(2) A 1-simplex has two orderings and two orientations: ordering and orientation are the same in this case. They are represented diagrammatically by an arrow from the first to the second vertex (compare the diagrams following 1.3). An orientation of a triangle is indicated by a circular arrow as in the above diagrams.

(3) For any permutation π of $0, \ldots, n$ the oriented simplexes $\sigma_n = (v^0 \ldots v^n)$ and $\tau_n = (v^{\pi(0)} \ldots v^{\pi(n)})$ satisfy $\sigma_n = \tau_n$ if π is even, and $\sigma_n = -\tau_n$ if π is odd.

Simplicial complexes

Simplicial complexes are the most general geometrical objects we study in this book. They include as special cases the graphs of Chapter 1 and the closed surfaces of Chapter 2, and they provide a good framework for developing an algebraic theory analogous to the cycles of Chapter 1. As before we impose 'intersection conditions' (compare 1.3, 2.1); now that we have the ideas of simplex and face at our disposal it is possible to state these very succinctly.

Definitions 3.14. A *simplicial complex* is a finite set K of simplexes in $\mathbb{R}^N$ with the following two properties:

(1) if $s \in K$ and $t < s$ (t is a face of s (3.9)) then $t \in K$
(2) intersection condition: if $s \in K$ and $t \in K$ then $s \cap t$ is either empty or else a face both of s and of t.

The *dimension* of K is the largest dimension of any simplex in K. The *underlying space* of K, denoted $|K|$, is the set of points in $\mathbb{R}^N$ which belong to at least one simplex of K. That is, $|K|$ is the union of the simplexes of K. Occasionally, if Σ is merely a set of simplexes we use $|\Sigma|$ to denote the union of the simplexes in Σ.

Notice that if $s \in K$ then all the faces of s are also elements of K. It may seem odd that we should explicitly include these faces in K since they are, as subsets of $\mathbb{R}^N$, contained in s: the faces of s do not contribute anything extra to the underlying space. The reason is to be found in the next chapter, where algebraic apparatus is introduced which requires the presence of all these faces. It is worth emphasizing at this point that a simplicial complex is a finite set of simplexes in $\mathbb{R}^N$ while the portion of $\mathbb{R}^N$ which these simplexes together occupy (i.e. the underlying space) is generally an infinite set of points in $\mathbb{R}^N$. (The underlying space $|K|$ is finite if and only if K has dimension 0.)

When *drawing simplicial complexes* it is sometimes helpful to indicate vertices by a heavy dot and to shade in 2-simplexes which are part of the complex. We shall occasionally make use of this in what follows. But for the majority of the diagrams we just use the convention that straight segments are 1-simplexes, intersections of segments (and no other points) are 0-simplexes, and all triangular regions in the diagram are in fact spanned by 2-simplexes. From now on the reader should assume this latter convention unless there is indication (usually by selective shading) to the contrary.

Examples 3.15. (1) A *graph* is a simplicial complex of dimension 1. This agrees with 1.3, although admittedly no explicit distinction is made there between the finite set of vertices and edges (the simplicial complex) and the, usually infinite, underlying space.

(2) The empty collection of simplexes is a simplicial complex whose underlying space is empty. This simplicial complex does not have a dimension, unless we make a special convention to cover it. A popular choice is -1.

(3) A closed surface M as defined in 2.1 consists of triangles (i.e. 2-simplexes) satisfying three conditions one of which is an intersection condition. If we let M^+ be the set of triangles, edges and vertices in M then M^+ is a simplicial complex of dimension 2: thus a closed surface is a simplicial complex provided we explicitly mention the vertices and edges as well as the triangles. In the future we shall assume this has been done. A closed surface is then a special kind of simplicial complex called 'pure': a simplicial complex K of dimension n is called *pure* if every simplex of K is a face of some n-simplex of K. An example of a non-pure simplicial complex is $K = \{(v^0 v^1), (v^0), (v^1), (v^2)\}$ where v^0, v^1, v^2 are distinct. Here dim $K = 1$ but (v^2) is not a face of any 1-simplex of K.

(4X) The three diagrams below do not represent simplicial complexes (Why?), but below them are diagrams for genuine simplicial complexes with, in each case, the same underlying space. The triples in brackets are the numbers of 0, 1 and 2-simplexes respectively.

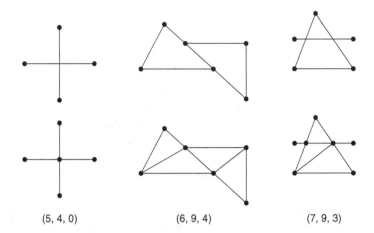

(5, 4, 0)   (6, 9, 4)   (7, 9, 3)

(5X) The set of all faces of an n-simplex s_n is a simplicial complex of dimension n (see 3.10(2)), which is denoted by $\bar{s}_n$. The set of all proper faces of s_n is a simplicial complex of dimension $n - 1$ (even for $n = 0$ if we adopt the convention of (2) above). It is denoted $\overset{\bullet}{s}_n$ and is called the *boundary* of s_n. Thus (compare 3.10(1)) the 'boundary of s_n' refers either to the set of proper faces of s_n or to the union of the proper faces of s_n. This will not cause difficulty.

(6XX) The conditions (1) and (2) of 3.14 are equivalent to (1) and the following condition (2)'.

(2)' If s and $t \in K$, where $s \neq t$, then the open simplexes $\overset{\circ}{s}$ and $\overset{\circ}{t}$ with the same vertices as s and t are disjoint.

It is not too hard to prove (1) & (2) $\Rightarrow$ (1) & (2)'. The proof of the converse implication is more difficult. One method is to use induction, assuming the result true whenever K contains only simplexes of dimension $< n$, and proving it when K contains only simplexes of dimension $\leqslant n$. The induction hypothesis then applies to the subset of K consisting of simplexes of dimension $< n$ and since open simplexes are disjoint we have $s \cap t = \overset{\circ}{s} \cap \overset{\circ}{t}$ for any $s, t \in K$. Hence by the induction hypothesis $s \cap t$ is a union of common faces of s and t. However it is also convex (see 3.8(4)) and it is not hard to show that a convex union of (closed) faces of a simplex is itself a face.

Definitions 3.16. (1) A *subcomplex* of a simplicial complex K is a subset L of K which is itself a simplicial complex. (This is so if and only if L has the property that

$$s \in L \quad \& \quad t < s \Rightarrow t \in L,$$

since 3.14(2) is automatically satisfied.) The subcomplex L is called *proper* if $L \neq K$.

(2) An *oriented simplicial complex* is a simplicial complex in which every simplex is provided with an orientation.

Thus every simplicial complex can certainly be oriented, usually in many different ways. This is not to be confused with the coherent orientation of a closed surface, as in 2.3, which is not always possible.

By orienting a simplicial complex each subcomplex becomes oriented in a natural way. Orientations of 1-simplexes and 2-simplexes will be indicated by arrows and circular arrows respectively.

Examples 3.17. (1X) If K_1 and K_2 are subcomplexes of K then so are $K_1 \cup K_2$ and $K_1 \cap K_2$. How about $K_1 \setminus K_2$?

(2XX) Let K_1 and K_2 be simplicial complexes in $\mathbb{R}^N$ and suppose that $|K_1| \cap |K_2| = |L|$, where L is a subcomplex both of K_1 and of K_2. Show that $K_1 \cup K_2$ is a simplicial complex. (There is no difficulty showing that (1) of 3.14 holds, and the only tricky case of (2) is when $s \in K_1$ and $t \in K_2$. It is probably easiest to prove that, supposing $s \neq t$, the open simplexes determined by s and t are disjoint, and then to use 3.15(6).)

(3) Let r be an integer ≥ 0. The *r-skeleton* of a simplicial complex K is the set of simplexes of K with dimension $\leq r$; it is denoted K^r. Clearly K^r is a subcomplex of K. The 1-skeleton K^1 of any simplicial complex of K is a graph.

(4) A simplicial complex K is called *connected* if K^1 is a connected graph, i.e. if there is a path along the edges of K^1 connecting any two given vertices. For any K, a *component* of K is a maximal connected subcomplex; thus K is the (disjoint) union of its components, and K is connected if and only if it has exactly one component.

Abstract simplicial complexes and realizations

A simplicial complex K is determined by (i) its set V of vertices and (ii) the subsets of V which span simplexes of K. This information can be presented abstractly, and the result is an 'abstract simplicial complex' which can be 'realized' to give K. The definitions are parallel to those given for graphs in Chapter 1, and again there is a theorem that realization is always possible. We shall make use of these ideas in the next section, in an attempt to clarify the meaning of certain diagrams representing simplicial complexes.

78   *Simplicial complexes*

Definition 3.18. An *abstract simplicial complex* is a pair $X = (V, S)$ where V is a finite set whose elements are called the *vertices* of X and S is a set of non-empty subsets of V. Each element $s \in S$ is called a *simplex* of X, and if $s \in S$ has precisely $n + 1$ elements ($n \geqslant 0$), s is called an *n*-simplex. (Thus an 'abstract' simplex is merely the set of its vertices.) S is required to satisfy the following two conditions:

(1) $v \in V \Rightarrow \{v\} \in S$
(2) $s \in S$, $t \subset s$, $t \neq \emptyset \Rightarrow t \in S$.

The *dimension* of X is, so long as $V \neq \emptyset$, the largest n for which S contains an *n*-simplex. (The dimension of $(\emptyset, \emptyset)$ is defined to be -1.)

Two abstract simplicial complexes $X = (V, S)$ and $X' = (V', S')$ are said to be *isomorphic* if there is a bijection $f : V \to V'$ with the property that

$$\{v^0, \ldots, v^n\} \in S \Leftrightarrow \{f(v^0), \ldots, f(v^n)\} \in S'.$$

It is clear that any simplicial complex K determines an abstract simplicial complex $X = (V, S)$: take V to be the set of vertices of K and let $\{v^0, \ldots, v^n\} \in S$ if and only if $(v^0 \ldots v^n) \in K$. Two simplicial complexes are called *isomorphic* if the abstract simplicial complexes they determine are isomorphic. A *realization* of an abstract simplicial complex X is a simplicial complex K whose corresponding abstract simplicial complex is isomorphic to X. Thus the vertices of K can be so labelled that a set of vertices of K spans an *n*-simplex of K if and only if the corresponding set of vertices of X is an *n*-simplex of X.

This idea generalizes 1.3 and the following theorem generalises 1.4. Readers interested only in two-dimensional simplicial complexes should substitute $n = 2$ throughout the theorem and proof. The proof could in any case safely be omitted on a first reading.

Theorem 3.19. *Every abstract simplicial complex of dimension n has a realization in $\mathbb{R}^{2n+1}$. (Note that any two realizations of the same abstract simplicial complex are by definition isomorphic.)*

Sketch of Proof Consider the 'rational normal curve'

$$C = \{(x, x^2, \ldots, x^{2n+1}) \in \mathbb{R}^{2n+1} : x \in \mathbb{R}\}.$$

I claim that any $2n + 2$ distinct points of C are *a*-independent. The proof is similar to that in 1.4 of the case $n = 1$: it amounts to checking that a certain determinant (a Vandermonde determinant) does not vanish.

Now let $X = (V, S)$ be an abstract simplicial complex and take points $P_1, \ldots, P_k$ on C, one for each element of V ($k = |V|$). If $s \in S$ then $s \subset V$ and $|s| \leq n + 1 < 2n + 2$ so the P's corresponding to elements of s are a-independent and span a ($|s| - 1$)-simplex. Let K be the set of such simplexes for all $s \in S$. It remains to prove that K is a simplicial complex for, if it is, then the underlying abstract simplicial complex is certainly X again.

Take two simplexes in K and suppose that their sets of vertices overlap. Then we can take them to be

$$(P_1 \ldots P_\alpha \ldots P_\beta) \text{ and } (P_\alpha \ldots P_\beta \ldots P_\gamma)$$

say, where $\gamma < 2n + 2$ since $\dim X = n$, so that $P_1, \ldots, P_\gamma$ are a-independent. Using the definition of a-independence and 3.2(2) it is not hard to show that these simplexes intersect precisely in $(P_\alpha \ldots P_\beta)$ which is a common face. If the simplexes have disjoint vertex sets it is equally easy to show that they are disjoint. □

Since orientation is defined entirely in terms of vertices, there is no difficulty in transferring the concept of orientation from simplicial complexes to abstract simplicial complexes.

Triangulations and diagrams of simplicial complexes

A subset U of $\mathbb{R}^N$ which is the underlying space (3.14) of some simplicial complex is called a *polyhedron*. Of course there will almost certainly be many simplicial complexes whose underlying space is a given polyhedron U (the only exceptions are when U is finite); any of these is called a *triangulation* of U.

Something more will be said about triangulations in Chapter 5. For the present we are concerned with diagrammatic representation, which we shall discuss by means of an example which sufficiently illustrates what is involved, and in particular clarifies the meaning of diagrams such as those in 2.2(3), (4) and preceding 2.6.

If we cut a torus along two circles and open it out as in 2.2(3) the result is a plane rectangle with opposite sides similarly marked in order to remind us that they came from the same place on the torus. Consider the three diagrams below (which occurred in Chapter 2). We want, if possible, to regard these as recipes or 'schemas' for the construction of simplicial complexes in which dots, edges and triangles represent 0-, 1- and 2-simplexes respectively. To do this, we first regard each of them as a schema for constructing an abstract simplicial complex $X = (V, S)$.

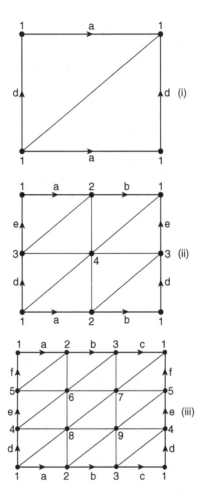

First label the vertices (dots) consistently, so that for example in any one diagram the beginning of a has the same label wherever it occurs. For notational convenience they have been labelled with numerals. Then take V to be the set of vertices, so that $V = \{1\}$ in (i), $V = \{1, 2, 3, 4\}$ in (ii) and $V = \{1, \ldots, 9\}$ in (iii).

Next define S as follows: a subset $s \subset V$ belongs to S if and only if
$s \in \{v\}$ for some $v \in V$
or s is the set of vertices of an edge in the diagram
or s is the set of vertices of a triangle in the diagram.

This defines an abstract simplicial complex $X = (V, S)$, which by 3.19 has a realization, but there are two questions to ask.

Qu. 1. Do edges and triangles in the diagram represent 1- and 2-simplexes of X?

Qu. 2. What connexion, if any, is there between a realization of X and the torus from which we started?

3.20. The answer to Qu. 1 is 'yes' if and only if any given edge in the diagram has differently labelled end-points (which implies that any given triangle has three differently labelled vertices).

This is clear from the definition of an n-simplex of X as a set of vertices with precisely $n + 1$ elements. Thus (i) does not define an abstract simplicial complex of the kind required. Each edge and each triangle of (i) gives only a 0-simplex of X, and a realization of X is just one point. On the other hand for (ii) and (iii) the answer to Qu. 1 is 'yes'. (In (ii), it does not matter that a and b have the same set of end-points; what matters is that the end-points of a are different and the end-points of b are different.)

As for Qu. 2, this demands a less formal answer. When (ii) or (iii) is realized two edges with the same vertices will automatically be realized in the same place, i.e. will give rise to the same 1-simplex, The question is: does the realization look anything like the original torus?

3.21. Suppose that two edges (or two triangles) of the diagram have the same vertices if and only if they came from the same position on the torus before it was cut open. Then we shall regard a realization of X as a triangulation of the torus.

This means that when X is realized only the necessary identifications take place, and no extra ones. (It follows that the underlying space of the realization is homeomorphic to the original torus.) When the X coming from (ii) is realized a great many extra identifications take place (for example a and b^{-1} are identified), and the reader may like to try drawing the result. [Hint: the bottom half of the diagram is a repetition of the top half, and can be discarded.]

Examples 3.22. (1X) Theorem 3.19 assures us that the X coming from (iii) can be realized in $\mathbb{R}^5$ (see also 2.16(5)) whereas the original torus was in $\mathbb{R}^3$. Can this X in fact be realized in $\mathbb{R}^3$? A harder question (for the answer see Gardner (1975,1978)) is whether the diagram below, which also gives a torus, can be realized in $\mathbb{R}^3$.

82   *Simplicial complexes*

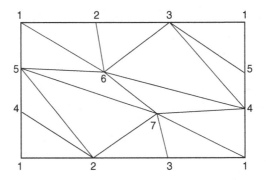

(2) In 2.6, (i) ensures that the answer to Qu. 1 is 'yes' while (ii) and (iii) relate to Qu. 2.

(3X)

This diagram represents a Möbius band (compare 2.2(6)) if and only if there are at least five triangles. Reversing one of the arrows it represents a cylinder (obtained by identifying the ends of a rectangle without a twist) if and only if there are at least six triangles.

(4X) Find a triangulation of the object obtained by identifying all three sides of a triangle, as in the diagram (this is sometimes referred to as a *dunce hat*).

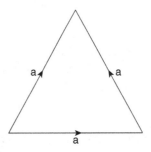

(5X) Cut a double torus (see 2.9) along suitable curves until it can be opened out flat, and hence find a triangulation. (You could use the method of 2.13 to check that your opened out diagram represents the double torus, which is orientable and has Euler characteristic -2.)

(6) Any diagram of the type exemplified by (i)–(iii) above can be subdivided into one in which the answer to Qu. 1 (Page 80) is 'Yes' and in which the answer

Triangulations and diagrams of simplicial complexes 83

to Qu. 2 is that the realization is homeomorphic to the torus (or whatever it was we started from). Two barycentric subdivisions are enough for (i), and one is enough for (ii). (Compare the discussion following 2.6.)

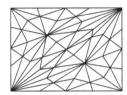 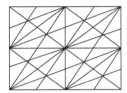

★(7) Diagrams such as (i)–(iii) above can be interpreted other than as recipes for the construction of abstract simplicial complexes. For example (i) can be interpreted as a covering of the torus by two triangles which are wrapped around the torus in such a way that each edge becomes a circle and all the

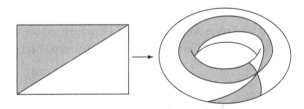

vertices land on the same point. Thus identifications take place on the boundaries of the triangles, but notice that there are no identifications of interior points. This kind of construction cannot be made with flat triangles, and the intersection condition 3.14(2) is hopelessly violated.

Likewise (ii) can be interpreted as eight triangles on the torus, and again the intersection condition is violated, though in this case each individual triangle lands on the torus without identifications. Of course with (iii) the intersection condition is satisfied.

These coverings by curved triangles which may have identifications on their boundaries and may violate the intersection condition are called *cell complexes*. It is possible to use much more general building blocks than triangles: for example the diagram below makes the torus into a cell complex with one 2-cell (the rectangle), two 1-cells (a and b) and one 0-cell (v).

84   *Simplicial complexes*

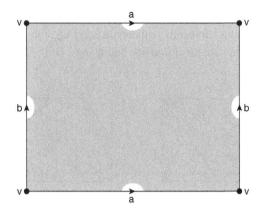

This is certainly a desirable situation: calculations with cells are easier than with simplexes because there are fewer of them. But you never get something for nothing: the definition of cell complex is technically much more complicated than that of simplicial complex, and requires considerable preparation of a 'general topology' kind. One reason for this is that flat triangles can be defined simply by a little linear algebra, whereas for curvy cells one cannot get away from the concept of continuous mappings between subsets of, at least, euclidean spaces. For more information on cell complexes, see for example the books of Sato or Prasolov *Elements of Combinatorial and Differential Topology* listed in the References. ★

Stars, joins and links

In this section we shall introduce some standard terms used in connexion with simplicial complexes, and state a few results for the reader to verify. The section is intended primarily as a source of examples from which the reader may develop a greater facility in working with and thinking about simplicial complexes. Only 3.23, 3.29(2) and 3.29(5) are of significance later. The first two are straightforward while the third is a fairly hard exercise giving the construction of 'stellar subdivisions' which are used in Chapter 5, itself mostly 'optional reading'. Enough is said in Chapter 5 to enable the reader who has not worked through 3.29(5) to follow the arguments, at any rate for the case of two-dimensional simplicial complexes.

Throughout, K is a simplicial complex.

3.23. Closure Let Σ be a set of simplexes of K. The *closure* of Σ (in K) is the smallest subcomplex of K which contains Σ, i.e. it is defined by

$$\overline{\Sigma} = \{s \in K \,:\, s < t \text{ for some } t \in \Sigma\}.$$

Other notations are Cl(Σ) and Cl(Σ, K). If $s \in K$ and $\Sigma = \{s\}$, then we shall write $\bar{s}$ rather than $\overline{\{s\}}$ for the closure of Σ. (Compare 3.15(5).) Note that $|\bar{s}| = s$ (3.14); indeed $|\Sigma| = |\overline{\Sigma}|$ for any Σ.

3.24. Star Let Σ be a set of simplexes of K. The *star* of Σ (in K) is defined by

$$\text{St}(\Sigma) = \text{St}(\Sigma, K) = \{s \in K : s > t \text{ for some } t \in \Sigma\}.$$

The *closed star* of Σ (in K) is the closure of the star of Σ, and is denoted $\overline{\text{St}}(\Sigma)$ or $\overline{\text{St}}(\Sigma, K)$. Thus

$$\overline{\text{St}}(\Sigma) = \{u \in K : \exists t \in \Sigma \text{ and } s \in K$$

$$\text{with } s > u \text{ and } s > t\}$$

and $\overline{\text{St}}(\Sigma)$ is a subcomplex of K. The most important example is $\Sigma = \{s\}$: we write $\text{St}(s)$ and $\overline{\text{St}}(s)$ rather than $\text{St}(\{s\})$ and $\overline{\text{St}}(\{s\})$.

Examples 3.25. (1X) Let s_0 be a 0-simplex of K. Then $s_0 \in \text{St}(\Sigma) \Leftrightarrow s_0 \in \Sigma$. It follows from this that $\text{St}(\Sigma)$ is a subcomplex of K if and only if, for each component K_1 of K, Σ contains either all or none of the 0-simplexes of K_1.

(2) Let K be the simplicial complex in the diagram, and let $\Sigma = \{s\}$. Then $\overline{\text{St}}(s) = \bar{t}$ where t is the 2-simplex. On the other hand $\text{St}(\bar{s}) = (\bar{t}\backslash\{v\}) \cup \{u, u'\}$. Thus $\overline{\text{St}}(\Sigma)$ and $\text{St}(\overline{\Sigma})$ may be sets neither of which contains the other.

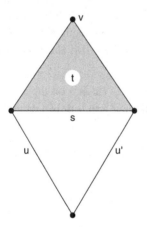

(3X) For $s \in K$,

$t \in \overline{\text{St}}(s) \Leftrightarrow s$ and t are both faces of the same simplex of K.

Thus the relation between s and t defined by $t \in \overline{\text{St}}(s)$ is symmetric (and reflexive). Is it transitive?

(4X) If $\Sigma_1 \subset \Sigma_2$ then $\text{St}(\Sigma_1) \subset \text{St}(\Sigma_2)$. Is the converse true?

3.26. Join Let $s_p = (v^0 \ldots v^p)$ and $s_q = (w^0 \ldots w^q)$ be simplexes in $\mathbb{R}^N$. They are called *joinable* if the $p + q + 2$ vectors $v^0, \ldots, v^p, w^0, \ldots, w^q$ are affinely independent, and in that case their *join* $s_p s_q$ is the $p + q + 1$-simplex $(v^0 \ldots v^p w^0 \ldots w^q)$. (If they are joinable then their join is undefined.) Thus $s_p s_q = s_q s_p$ when either side is defined.

Two simplicial complexes K and L in $\mathbb{R}^N$ are called *joinable* if, for each $s \in K$ and $t \in L$, s and t are joinable, and if any two such joins, st and $s't'$, meet, if at all, in a common face of st and $s't'$. The *join* KL of K and L is then the simplicial complex

$$KL = K \cup L \cup \{st \; : \; s \in K \text{ and } t \in L\}.$$

Examples 3.27. (1XX) Let $a \in s_p$ and $b \in s_q$; then the segment joining a and b consists of points $\lambda a + (1 - \lambda)b$ for $0 \leqslant \lambda \leqslant 1$. We have: s_p and s_q are joinable if and only if any two such segments intersect at most in an end-point, and in that case

$$s_p s_q = \{\lambda a + (1 - \lambda)b \; : \; a \in s_p, \quad b \in s_q, \quad 0 \leqslant \lambda \leqslant 1\}.$$

Likewise two simplicial complexes K and L are joinable if and only if the segments $\{\lambda a + (1 - \lambda)b \; : \; 0 \leqslant \lambda \leqslant 1\}$ are disjoint except for end-points for all $a \in |K|$ and $b \in |L|$, and $|KL|$ is then the union of all such segments. It follows that if $|K| = |K_1|$ and $|L| = |L_1|$ and K and L are joinable, then so are K_1 and L_1 and $|KL| = |K_1 L_1|$.

(2) The simplicial complexes (in $\mathbb{R}^2$) on the left are not joinable, whereas those in the centre are, and their join is drawn on the right.

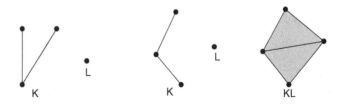

(3) Any K and the empty simplicial complex $\emptyset$ are joinable, and $K\emptyset = K$.

(4) If L consists of a single 0-simplex (v), and K and L are joinable, then KL is also called the *cone* on K with vertex v. We shall meet this idea again in 6.8.

(5X) Writing α_p for the number of p-simplexes of a simplicial complex, and supposing KL is defined,

$$\alpha_p(KL) = \alpha_p(K) + \left(\sum_{q=0}^{p-1} \alpha_{p-q-1}(K)\alpha_q(L)\right) + \alpha_p(L)$$

(the middle sum being absent when $p = 0$). In particular let K and L be 0-dimensional, consisting of m and n 0-simplexes respectively. Then KL is a graph with cyclomatic number $(m-1)(n-1)$. (Clearly KL is bipartite (1.21(5)).)

(6X) If the simplexes in 3.26 are regarded as oriented, then the orientation of $\sigma_p\sigma_q$ can be defined by the ordering $(v^0 \ldots v^p w^0 \ldots w^q)$. In that case,

$$\sigma_p\sigma_q = (-1)^{(p+1)(q+1)}\sigma_q\sigma_p.$$

3.28. Link Let $s \in K$. The *link* of s in K is defined by

$$\text{Lk}(s) = \text{Lk}(s, K) = \{t \in K : st \in K\}.$$

Recall that $st \in K$ means (i) s and t are joinable, and (ii) the join st belongs to K. It follows that $\text{Lk}(s)$ is a subcomplex of K.

Examples 3.29. (1) In the case of the simplicial complex of 3.25(2), $\text{Lk}(t) = \emptyset$, $\text{Lk}(s) = \{v\}$, $\text{Lk}(v) = \bar{s}$.

(2X) Let K be a simplicial complex of dimension 2. Then the link of a 0-simplex v^0 consists of all the 0-simplexes v^1 for which $(v^0 v^1) \in K$ and all the 1-simplexes $(v^1 v^2)$ for which $(v^0 v^1 v^2) \in K$. The link of a 1-simplex $(v^0 v^1)$ consists of all 0-simplexes v^2 for which $(v^0 v^1 v^2) \in K$.

If we suppose that the link of every 0-simplex is a simple closed polygon, then it follows that the link of every 1-simplex is two 0-simplexes and that K is pure (every simplex is a face of a 2-simplex). Thus K consists precisely of the triangles of K together with their edges and vertices. Hence the definition of a closed surface given in 2.1 can be rewritten:

A closed surface *is a connected two-dimensional simplicial complex in which the link of every 0-simplex is a simple closed polygon.*

(3X) Let s_0 be a 0-simplex of K. Then

$$\text{Lk}(s_0) = \overline{\text{St}}(s_0)\setminus\text{St}(s_0).$$

Is this ever true of an n-simplex for $n > 0$?

(4X) Let $s \in K$. Then $\bar{s}$ and Lk(s) are joinable, and $\bar{s}$Lk(s) = $\overline{\text{St}}(s)$. In particular if s is a 0-simplex s_0, then $\overline{\text{St}}(s_0)$ is the cone on Lk(s_0) with vertex s_0.

(5XX) Let $s \in K$ and let v be an interior point of s, i.e. a point of s not on a proper face of s. Let v also denote the simplicial complex whose only simplex is (v). Then v and $\overset{\bullet}{s}$ (the set of proper faces of s, 3.15(5)) are joinable, and their join $v\overset{\bullet}{s}$ is joinable to Lk(s, K). Further, $|v\overset{\bullet}{s}| = s$, and $|(v\overset{\bullet}{s})\text{Lk}(s, K)| = |\overline{\text{St}}(s, K)|$. (3.8(5) is relevant to the proof of $|v\overset{\bullet}{s}| = s$; see also 3.27(1).) Finally, $(K \setminus \overline{\text{St}}(s, K)) \cup (v\overset{\bullet}{s})\text{Lk}(s, K)$ is a simplicial complex with the same underlying space as K. It is said to be obtained from K by *starring s at v*. Repeated application of this procedure yields a *stellar subdivision* of K. Some examples of starring are drawn below.

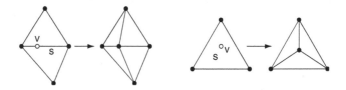

Collapsing

The technique of collapsing is a simple geometrical idea which will be exploited in Chapter 6. Since, however, it does not assume any of the theory of homology groups, we shall introduce the idea here and give one or two mild applications.

Definitions 3.30. Let K be a simplicial complex. A *principal simplex* of K is a simplex of K which is not a proper face of any simplex of K. For any $s \in K$, a *free face* of s is a proper face t of s such that t is not a proper face of any simplex of K besides s. (Thus a free face of s has dimension dim $s - 1$ and if s has a free face then s must be principal.)

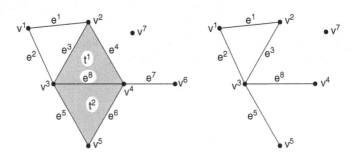

In the left-hand diagram, the principal simplexes are:

$$v^7; e^1, e^2, e^7; t^1, t^2.$$

Of these, only e^7, t^1, t^2 have free faces, namely v^6 for e^7, e^3 and e^4 for t^1, and e^5 and e^6 for t^2.

Suppose that K has a principal simplex s which has a free face t. Then the subset K_1 of K obtained by removing s and t is a subcomplex of K, and is said to be obtained from K by an *elementary collapse* (which we describe as being a *collapse of s across t*). The notation is $K \searrow^e K_1$. The right-hand diagram above shows the result of three elementary collapses (t^1 across e^4, t^2 across e^6 and e^7 across v^6); two further elementary collapses, of e^5 across v^5 and e^8 across v^4, now become possible. A sequence $K \searrow^e K_1 \searrow^e K_2 \searrow^e \ldots \searrow^e K_m$ of elementary collapses is called a *collapse*, and is denoted by $K \searrow K_m$. (A special case is the empty sequence by which $K \searrow K$!) If K_m consists of a single vertex we say that K *collapses to a point*.

This concept has already been encountered for graphs in 1.12(2): a graph collapses to a point if and only if it is a tree.

Let $\mathscr{D}$ be the plane region (a *disk*) enclosed by a simple closed polygon (see 1.25), and let D be a triangulation of $\mathscr{D}$, i.e. a simplicial complex (of dimension two) such that $|D| = \mathscr{D}$. Our object is to prove that D collapses to a point. A naïve argument would run: 'Collapse one triangle of D across a free face on the boundary polygon. What is left is a triangulation D' of a region $\mathscr{D}'$ enclosed by a simple closed polygon, and by induction $D' \searrow$ point, Q.E.D.' Unfortunately, this will not do, as can be seen by collapsing s across t in the diagram below: what is left is not enclosed by a simple closed polygon. The trouble is that s meets the boundary polygon in an edge and an isolated vertex.

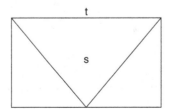

Lemma 3.31. *Let us denote the simple closed polygon enclosing $\mathscr{D}$ by ∂D. Suppose that D contains at least two triangles; then D contains at least two triangles meeting ∂D in an arc, i.e. in an edge or two edges. (See 3.22(3) for*

a picture which, on removing the arrows, gives an example where D contains exactly two such triangles.)

Proof The proof is by induction on $\alpha_2(D)$ the number of triangles of D: clearly it is true when $\alpha_2 = 2$ so suppose it holds when $\alpha_2 < n$ ($n \geqslant 3$) and let $\alpha_2(D) = n$. Let s be a triangle of D meeting ∂D in at least an edge – clearly D must contain at least two such triangles. Then $s \cap \partial D$ is

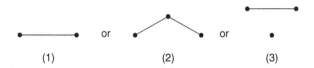

If there are no triangles of type (3) in D then all choices of s meet ∂D in an arc, so the result is proved. If there is a triangle s of type (3), proceed as follows (see the diagram below). Label the vertices of ∂D by $P_1, P_2, P_3, \ldots, P_m$ where s has vertices $P_1 P_2 P_i$ for some i, $4 \leqslant i \leqslant m-1$. Then $P_2 P_3 \ldots P_i P_2$ and $P_1 P_i P_{i+1} \ldots P_m P_1$ are simple closed polygons enclosing plane regions $\mathscr{D}'$, $\mathscr{D}'' = 1$ triangulated by D', D'', say. If $\alpha_2(D') = \alpha_2(D'') = 1$, then the unique triangles in D', D'' meet ∂D in an arc. If $\alpha_2(D') > 1, \alpha_2(D'') = 1$, then by the induction hypothesis choose two triangles of D' meeting $\partial D'$ in an arc. At most one of these can contain the edge $P_2 P_i$; the other[†], together with the unique triangle in D'', both meet ∂D in an arc. The other cases are treated similarly; in every case D contains at least two triangles meeting ∂D in an arc.

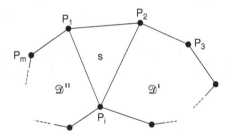

□

Theorem 3.32. *With the above notation, any triangulation D of $\mathscr{D}$ collapses to a point.*

[†] Notice that the proof would break down at this point if we were trying to prove the weaker assertion that D always contains at least one triangle meeting ∂D in an arc. This is a not uncommon feature of proofs by induction.

Collapsing

Proof Clearly the theorem holds when $\alpha_2(D) = 1$, the collapse then being

Suppose the theorem is true when $\alpha_2(D) < n$ $(n \geqslant 2)$ and let $\alpha_2(D) = n$. By the lemma choose a triangle s of D meeting ∂D in an arc, and collapse as follows:

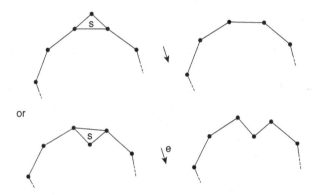

In either case $D \searrow D_1$ where D_1 is a triangulation of the plane region bounded by a simple closed polygon, and $\alpha_2(D_1) = n - 1$. Thus $D \searrow D_1 \searrow$ point, by the induction hypothesis. □

The *Euler characteristic* of a simplicial complex K of dimension n is $\chi(K) = \alpha_0(K) - \alpha_1(K) + \alpha_2(K) - \cdots + (-1)^n \alpha_n(K)$, α_i being the number of simplexes of dimension i in K, $i = 0, \ldots, n$. (Compare 2.11.) Clearly if $K \searrow L$ then $\chi(K) = \chi(L)$ (consider the effect of an elementary collapse). Hence:

Corollary 3.33. *If D is a triangulation of the plane region $\mathscr{D}$ enclosed by a simple closed polygon, then $\chi(D) = 1$.* (Compare the proof of 2.13.) □

There is another consequence of 3.31 which will be useful later.

Theorem 3.34. *Let D be a triangulation of the plane region $\mathscr{D}$ enclosed by a simple closed polygon ∂D, and let the simplicial complex K be obtained from D by deleting one 2-simplex. Then $K \searrow \partial D$.*

Proof There is nothing to prove if $\alpha_2(D) = 1$, for then $K = \partial D$ and the empty sequence of elementary collapses will work! Suppose the result is true when $\alpha_2 < n$ ($n \geqslant 2$) and that $\alpha_2(D) = n$. By the lemma (3.31) K has at least one triangle s meeting ∂D in an arc. Collapse as in the proof of 3.32; let this take D to D' and K to K'. Then D' is a triangulation of a disk containing $n - 1$ triangles, and K' is obtained from D' by removing a single triangle s; hence by the induction hypothesis $K' \searrow \partial D'$. Now none of the elementary collapses comprising this collapse can be across a face of s, since the face or faces of s in K' are still present at the end, in $\partial D'$. Hence $K = K' \cup \bar{s} \searrow \partial D' \cup \bar{s}$ where $\bar{s}$ denotes, as usual, the set of all faces of s including s itself. Since $\partial D' \cup \bar{s} \searrow \partial D$ this completes the induction. □

Examples 3.35. (1) This result can be used to given an alternative proof of 2.13 without using $\chi(D) = 1$. For removing a single triangle from a closed surface gives a simplicial complex which collapses to the one-dimensional subcomplex which appears, in a polygonal representation of the surface in the plane, as the boundary of the polygon. The Euler characteristic of this graph is easy to calculate, and the removal of a triangle at the beginning just lowers the Euler characteristic by 1. For example the triangulation of a torus in 2.2(3) collapses, after removal of a triangle, to the graph below which has Euler characteristic -1. Thus the original simplicial complex has Euler characteristic 0.

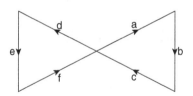

(2) The triangulations of a cylinder and a Möbius band in 3.22(3) collapse to simple closed polygons (sometimes called *spines* of the original simplicial complexes).

(3X) Let M be a closed surface and let K be a subcomplex of M which does not contain all the triangles of M. Supposing that K contains at least one triangle, then it contains at least one triangle with a free face, and therefore K collapses to a subcomplex with fewer triangles and so by induction to a graph. (To see that K must have a triangle with a free face, use 2.2(5) to show that otherwise K would have to contain all the triangles of M.)

(4) It is perhaps worth noting that 3.32 does not generalize to higher dimensions. A simplicial complex in $\mathbb{R}^3$ whose underlying space is homeomorphic to the unit sphere $\{(x, y, z) \in \mathbb{R}^3 : x^2 + y^2 + z^2 = 1\}$ in $\mathbb{R}^3$ will enclose a region known as a *3-ball*. It is possible to find such a simplicial complex and a

triangulation of the resulting 3-ball which does not collapse to a point. A famous example is called 'Bing's house with two rooms'. See Bing (1964).

★ Appendix on orientation

The passage from simplexes to oriented simplexes is an essential one for setting up the homology theory described in this book. The concept of orientation arises also in the study of real vector spaces, and in this appendix we shall relate these two ideas as well as proving the basic property of orientations – that there are always exactly two of them.

Let V be a real vector space of dimension $n \geqslant 1$ provided with a definite basis, called the *preferred basis*, and suppose that $e^1, \ldots, e^n$ and $f^1, \ldots, f^n$ are *ordered bases* of V, that is they are bases in which the vectors come in a prescribed order. (Such a basis is sometimes called an *n-frame*.) We want to compare the 'orientations' of these two bases.

Take for example $V = \mathbb{R}^2$. Then geometrically e^1, e^2 and f^1, f^2 have the same orientation if and only if they are both 'clockwise' or both 'anticlockwise' – more precisely this means that the rotations from e^1 to e^2 and from f^1 to f^2, in each case through an angle $< 180°$, are both in the same direction. Thus in the diagram below the first and second bases have the same orientation, not shared by the third basis. When $V = \mathbb{R}^3$ then geometrically we require that the two bases should both be 'right-handed' or both 'left-handed'. Thus for example for any basis e^1, e^2, e^3 and e^2, e^1, e^3 have opposite orientation while e^1, e^2, e^3 and $-e^2, e^1, e^3$ have the same orientation.

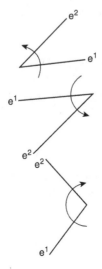

Now we ask the question: can we find a continuous family[†] of ordered bases $u^1(t), \ldots, u^n(t)$, depending on a parameter $t \in [0, 1]$, such that for $t = 0$ this basis is $e^1, \ldots, e^n$ and for $t = 1$ it is $f^1, \ldots, f^n$? (The important thing is that for every $t \in [0, 1]$ we must get a basis; we can turn e^1, e^2 (an ordered basis of $\mathbb{R}^2$) into e^2, e^1 by the family $(1-t)e^1 + te^2$, $te^1 + (1-t)e^2$, but for $t = \frac{1}{2}$ these vectors are not linearly independent.) The reader can probably convince himself that for $V = \mathbb{R}^2$ and $V = \mathbb{R}^3$ such a family exists if and only if the two ordered bases have the same orientation in the intuitive sense described above.

Any linear isomorphism $V \to V$ determines, relative to the preferred basis in each copy of V, a nonsingular real $n \times n$ matrix. The isomorphism is called *orientation preserving* if the determinant of this matrix is positive, and *orientation reversing* if it is negative. Note that this definition does not in reality depend on the choice of preferred basis, for if A is the matrix relative to the preferred basis then the matrix relative to any other basis is of the form $P^{-1}AP$ for some nonsingular P, and this has the same determinant as A. I now claim the following.

Theorem 3.36. *Let $e^1, \ldots, e^n$ and $f^1, \ldots, f^n$ be ordered bases of V. The following statements are equivalent:*

(i) *There exists a continuous family of ordered bases taking $e^1, \ldots, e^n$ to $f^1, \ldots, f^n$.*
(ii) *The unique isomorphism $V \to V$ taking e^i to f^i for $i = 1, \ldots, n$ is orientation preserving.*

When these hold we say that the two bases have the same orientation, *otherwise that they have* opposite orientation.

The theorem will be proved at the end of this appendix. It says that ordered bases can be divided into two orientation classes in such a way that two bases in the same class can be connected by a continuous family and two bases in different classes cannot. The classes can be distinguished by means of a determinant.

Examples 3.37. (1X) $e^1, \ldots, e^n$ and $e^{\pi(1)}, \ldots, e^{\pi(n)}$, where π is a permutation of the set $\{1, \ldots, n\}$ have the same orientation if and only if π is an even permutation. It is enough to check that if π is a transposition then the two bases

---

[†] Write $u^j(t) = (a_{1j}(t), \ldots, a_{nj}(t))$, where these are the coordinates of $u^j(t)$ relative to the preferred basis. Then we require that the n^2 functions $a_{ij}(t)$ should be continuous. (This definition is in fact independent of the choice of preferred basis.)

have opposite orientation, and this is most easily proved by taking (as we may) $e^1, \ldots, e^n$ itself to be the preferred basis.

(2X) $e^1, \ldots, e^n$ and $-e^1, \ldots, -e^n$ have the same orientation if and only if n is even.

Now let us turn to *orientation of simplexes*. Let $(v^0 \ldots v^n)$ and $(v^{\pi(0)} \ldots v^{\pi(n)})$ be two ordered simplexes, where π is a permutation of $\{0, \ldots, n\}$. The vectors $v^0, \ldots, v^n$ lie in an affine subspace U^A of dimension n in $\mathbb{R}^N$, while the vectors $v^i - v^j$ lie in a linear space U of dimension n parallel to U^A. Consider the following two bases of U.

$$v^1 - v^0, \ldots, v^n - v^0$$

and $v^{\pi(1)} - v^{\pi(0)}, \ldots, v^{\pi(n)} - v^{\pi(0)}$.

Definition 3.38. $(v^0 \ldots v^n)$ and $(v^{\pi(0)} \ldots v^{\pi(n)})$ have the *same* or *opposite orientation* according as these two bases of U have the same or opposite orientation.

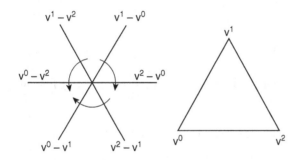

Examples 3.39. $(v^0 v^1 v^2) = (v^1 v^2 v^0)$ (same orientation)
$(v^0 v^1 v^2) = -(v^2 v^1 v^0)$ (opposite orientation)

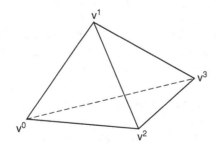

$v^1 - v^0, v^2 - v^0, v^3 - v^0$ is 'right-handed'; $v^3 - v^2, v^1 - v^2, v^0 - v^2$ is 'left-handed'.
Hence $(v^0 v^1 v^2 v^3) = -(v^2 v^3 v^1 v^0)$.

Proposition 3.40. $(v^0 \ldots v^n)$ *and* $(v^{\pi(0)} \ldots v^{\pi(n)})$ *have the same orientation if and only if π is an even permutation.*

Proof It is enough to check that if π is a transposition then the two simplexes have opposite orientation. Assuming that π is a transposition leaving 0 fixed the result follows from 3.37(1). Assuming on the other hand that π transposes 0 and i for some $1 \leqslant i \leqslant n$, the matrix, referred to the first basis above, of the isomorphism taking the first basis to the second, has the following form.

$$\begin{pmatrix} 1 & & & & & & & & \\ & 1 & & & & & & & \\ & & \ddots & & & & & & \\ & & & 1 & & & & & \\ -1 & -1 & \cdots & -1 & -1 & -1 & \cdots & -1 \\ & & & & & 1 & & & \\ & & & & & & \ddots & & \\ & & & & & & & 1 \end{pmatrix}$$

Here the blank entries are all zero, and the line of -1s is the ith row. This clearly has determinant -1. $\square$

To conclude this appendix, here is a proof of Theorem 3.36.

Proof of 3.36 Any basis of V determines a (real) nonsingular $n \times n$ matrix whose jth column consists of the coordinates of the jth basis vector relative to the preferred basis. If E and F are the matrices determined in this way by the given bases, then the matrix, relative to the preferred basis, of the unique isomorphism taking e^i to f^i for $i = 1, \ldots, n$ is FE^{-1}. Thus statements (i) and (ii) of the theorem are respectively equivalent to

(i)' there is a continuous family of nonsingular $n \times n$ matrices taking E to F;
and (ii)' the determinants $|E|$ and $|F|$ have the same sign.

The implication (i)' $\Rightarrow$ (ii)' is a straightforward consequence of the fact that the determinant of a matrix, being a polynomial in the entries of the matrix, is a continuous function of the entries. Thus if the general matrix of the family is A_t (so that $A_0 = E$ and $A_1 = F$) then $|A_t|$, being continuous and non-zero for $0 \leqslant t \leqslant 1$, must have the same sign for $t = 0$ and $t = 1$.

Appendix on orientation

To prove (ii)′ ⇒ (i)′ we must exhibit a continuous family taking E to F. It is enough to construct a continuous family taking E to one of the two diagonal matrices diag $(1, 1, \ldots, 1, \pm 1)$. The family we shall construct will be made up of a sequence of families, corresponding to a sequence of operations which reduce E to one of these two matrices. Let A_0 be any nonsingular $n \times n$ matrix; consider the following two operations.

I. Multiply all the elements of one row (or column) of A_0 by $\lambda > 0$.
II. Add to one row (or column) of A_0 any multiple μ of a different row (or, respectively, column).

Note that neither of these affects the sign of $|A_0|$. Suppose A_1 is obtained from A_0 by an operation I. Then the general member of a continuous family taking A_0 to A_1 is A_t, where A_t is obtained from A_0 by an operation I, replacing λ by $1 - t + \lambda t$. Similarly an operation II can be achieved via a continuous family.

Now using operations I and II any nonsingular $n \times n$ matrix $E = (a_{ij})$ can be reduced to a diagonal matrix with 1's and -1's on the diagonal. We show how to turn E into a matrix of the form

$$\begin{pmatrix} \pm 1 & 0 & \cdots & 0 \\ 0 & & & \\ \vdots & & \boxed{} & \\ 0 & & & \end{pmatrix}$$

where the box contains an $(n - 1) \times (n - 1)$ nonsingular matrix; the proof then proceeds by induction. In fact, if $a_{11} = 0$ then choose a non-zero entry in the first row or column (they can't all be zero); say this is the entry a_{i1} in the ith row. Then by an operation II, with $\mu = 1/a_{i1}$, we obtain a matrix with top left-hand entry 1. If $a_{11} \neq 0$ multiply the first row by $1/|a_{11}|$ (an operation I), obtaining a matrix with top left-hand entry ± 1. Now the other entries in the first row and then the first column can easily be made zero by operations II.

Next, two -1's on the diagonal can be turned into $+1$'s:

$$\begin{pmatrix} -1 & 0 \\ 0 & -1 \end{pmatrix} \to \begin{pmatrix} -1 & -1 \\ 0 & -1 \end{pmatrix} \to \begin{pmatrix} 1 & -1 \\ 2 & -1 \end{pmatrix} \to \begin{pmatrix} 1 & -1 \\ 0 & 1 \end{pmatrix} \to \begin{pmatrix} 1 & 0 \\ 0 & 1 \end{pmatrix}.$$

Each operation is of type II. Similarly a 1 and a -1 can be interchanged by a sequence of operations of type II. This completes the proof that every E can be reduced to one of diag $(1, 1, \ldots, 1, \pm 1)$. □

An alternative statement of the theorem is that the general linear group $GL(n, \mathbb{R})$ contains precisely two path components for any $n \geqslant 1$. The proof above constructs a polygonal path joining any two matrices in this group with determinants of equal sign, regarding $GL(n, \mathbb{R})$ as being contained in $\mathbb{R}^{n^2}$ in the natural way. ★

4
Homology groups

In this chapter we shall formalize the ideas sketched in the Introduction and define homology groups for an arbitrary oriented simplicial complex K. There is one group, $H_p(K)$, in each dimension p with $0 \leqslant p \leqslant \dim K$; the group $H_p(K)$ measures, roughly speaking, the number of 'independent p-dimensional holes' in K. If K is an oriented graph then $H_1(K)$ is isomorphic to the group $Z_1(K)$ of 1-cycles on K (see 1.19).

In order to read this chapter it is necessary to know about quotient groups and exact sequences, and to know a little about presentations. All these are explained in the appendix to the book.

Chain groups and boundary homomorphisms

The chain groups associated with an oriented simplicial complex are exactly analogous to those associated with an oriented graph (1.17), and are defined as follows.

Definition 4.1. Let K be an oriented simplicial complex of dimension n, and let α_p be the number of p-simplexes of K. For $0 \leqslant p \leqslant n$, let $\sigma_p^1, \ldots, \sigma_p^{\alpha_p}$ be the oriented p-simplexes of K; for such p the pth *chain group* of K, denoted $C_p(K)$, is the free abelian group on the set $\{\sigma_p^1, \ldots, \sigma_p^{\alpha_p}\}$ (A.11). Thus an element of $C_p(K)$ is a linear combination

$$\lambda_1 \sigma_p^1 + \cdots + \lambda_{\alpha_p} \sigma_p^{\alpha_p}$$

with the λ's integers. This is called a *p-chain* on K and two p-chains are added by adding corresponding coefficients. When writing down p-chains it is customary to omit p-simplexes whose coefficient is zero – unless they are all zero, when we just write 0. We also write σ for 1σ. For $p > n$ or $p < 0$ the group $C_p(K)$ is defined to be 0. (Notice that $C_p(\emptyset) = 0$ for all p, since $\emptyset$ has no simplexes at all.)

We shall shortly introduce a 'boundary homomorphism' which generalizes that for graphs (1.19). Recall that the boundary $\partial(v^0 v^1)$ of an oriented 1-simplex $(v^0 v^1)$ is the 0-chain $v^1 - v^0$. Now consider an oriented 2-simplex $(v^0 v^1 v^2)$. It is certainly desirable that the boundary should be the 1-chain

$$\partial(v^0 v^1 v^2) = (v^1 v^2) + (v^2 v^0) + (v^0 v^1).$$

However, if the 2-simplex belongs to an oriented simplicial complex K then the three edges of the triangle will also have specific orientations as elements of K, and it might be that, for example, the bottom edge has orientation $(v^2 v^1)$. In that case $(v^1 v^2) \notin K$ so that the above formula does not in fact give a 1-chain: we should really write

$$\partial(v^0 v^1 v^2) = -(v^2 v^1) + \cdots$$

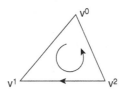

Alternatively – and this is the course we adopt – we can make the convention that

$(v^1 v^2)$ means the 1-chain $(v^1 v^2)$ if $(v^1 v^2) \in K$

and $(v^1 v^2)$ means the 1-chain $-(v^2 v^1)$ if $(v^2 v^1) \in K$.

Then the above formula for $\partial(v^0 v^1 v^2)$ always gives a 1-chain. In addition, this convention enables us to preserve the natural order of indices and write

$$\partial(v^0 v^1 v^2) = (v^1 v^2) - (v^0 v^2) + (v^0 v^1).$$

More generally, the convention is as follows.

Convention 4.2. Let $(v^0 \ldots v^p)$ be an oriented p-simplex $(p \geq 1)$. Let K be an oriented simplicial complex and suppose that the oriented simplex $\sigma = -(v^0 \ldots v^p)$, i.e. $\sigma = (v^{\pi(0)} \ldots v^{\pi(p)})$ for an odd permutation π, belongs to

K. Then we make the convention that $(v^0 \ldots v^p)$ means the p-chain $-\sigma$, i.e. the p-chain on K in which the coefficient of σ is -1 and every other coefficient is 0. Thus for an odd permutation π

$$(v^0 \ldots v^p) = -(v^{\pi(0)} \ldots v^{\pi(p)})$$

both as oriented simplexes (see 3.12) and as p-chains on K.

Definition 4.3. Let $\sigma = (v^0 \ldots v^p)$ be an oriented p-simplex of K (K as in 4.1) for some $p > 0$. The *boundary* of σ is the $(p-1)$-chain

$$\partial \sigma = \partial_p \sigma = \sum_{i=0}^{p} (-1)^i (v^0 \ldots \widehat{v^i} \ldots v^p)$$

where the hat over v^i indicates that it is omitted. Note that we are using the convention 4.2 here, for the face of σ with v^i omitted may or may not have the right orientation to be a $(p-1)$-simplex of K. For $p = 0$, $\partial_p \sigma$ is defined to be zero. The *boundary homomorphism* $\partial = \partial_p : C_p(K) \to C_{p-1}(K)$ is defined by

$$\partial_p(\Sigma \lambda_i \sigma_p^i) = \Sigma \lambda_i \partial_p(\sigma_p^i) \quad \text{(compare A.12)}$$

for $0 \leqslant p \leqslant n$, and defined to be the trivial homomorphism otherwise.

As an example, consider the two coherently oriented triangles in the diagram on the left.

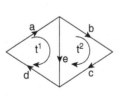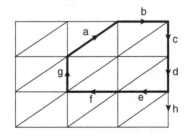

We have $\partial(t^1 + t^2) = (a + e + d) + (b + c - e) = a + b + c + d$, corresponding to the geometrical idea that the boundary of the whole area is the polygon $a+b+c+d$. Notice that $\partial(a+b+c+d) = 0$. Similarly, orienting all the triangles in the right-hand diagram clockwise, the boundary of the sum of the seven triangles enclosed by the heavy line is $a+b+c+d+e+f+g$, which is a 1-chain with boundary 0. If we make the usual identifications to turn the simplicial complex on the right into a torus (see 2.2(3)) then $\partial(c+d+h) = 0$, but it can be shown quite easily that there is no 2-chain with boundary $c+d+h$: the simple closed polygon $c+d+h$ goes round a hole in the torus and does not bound any region. Thus, on the torus, $a + \cdots + g$ bounds a region (is 'homologous to zero' as we shall say shortly) whereas $c+d+h$ does not. The homology groups we study pick out the simple closed polygons – more generally the 1-cycles – which do not bound any 2-chain by taking the group of 1-cycles and factoring out by the subgroup consisting of those which do bound 2-chains.

Remark 4.4. It is far from obvious that the above definition of $\partial \sigma$ makes sense – i.e. that it depends only on the oriented simplex σ and not on the particular ordering of vertices given. For instance, $(v^0 v^1 v^2) = (v^1 v^2 v^0)$ (as oriented simplexes), and

$$\partial(v^0 v^1 v^2) = (v^1 v^2) - (v^0 v^2) + (v^0 v^1);$$
$$\partial(v^1 v^2 v^0) = (v^2 v^0) - (v^1 v^0) + (v^1 v^2).$$

These expressions are indeed equal (bearing 4.2 in mind), but perhaps some general reassurance is necessary, (The reassurance which follows could be omitted on a first reading.) We shall also check that the formula is compatible with the notation (3.12) $-\sigma$ for the same simplex as σ but with opposite orientation. Both these things are contained in the statement

$$\partial\left(v^{\pi(0)} \ldots v^{\pi(n)}\right) = \pm \partial(v^0 \ldots v^n)$$

where the sign is $+$ or $-$ according as π is even or odd, and it is enough to verify this when π is a transposition (which will be odd).

Suppose, then, that $n \geqslant 1$ and π transposes i and j where $i < j$. Consider the face of σ not containing the vertex v^j. This occurs in the formula for $\partial(v^0 \ldots v^n)$ as

$$(-1)^j (v^0 \ldots v^i \ldots \widehat{v^j} \ldots v^n) = (-1)^j \tau$$

say and in the formula for $\partial(v^{\pi(0)} \ldots v^{\pi(n)})$ as

$$(-1)^i (v^0 \ldots \widehat{v^j}, \ldots v^i \ldots v^n) = (-1)^i \tau' \quad \text{say.}$$

Now moving v^i to the left $j - i - 1$ places turns τ' into τ. This amounts to effecting $j - i - 1$ transpositions, so

$$(-1)^i \tau' = (-1)^i (-1)^{j-i-1} \tau = -(-1)^j \tau.$$

Similarly the face of σ not containing v^i occurs in the two formulae with opposite sign, while a face of σ containing both v^i and v^j occurs with the same power of -1 but with v^i and v^j interchanged – hence altogether with opposite sign. This completes the verification.

Definition 4.5. The *augmentation*

$$\varepsilon : C_0(K) \to \mathbb{Z}$$

is the homomorphism defined by

$$\varepsilon(\Sigma \lambda_i \sigma_0^i) = \Sigma \lambda_i$$

(compare 1.22).

The sequences (not in general exact)

$$\ldots 0 \to 0 \to C_n(K) \xrightarrow{\partial} C_{n-1}(K) \to \cdots \to C_1(K) \xrightarrow{\partial} C_0(K) \to 0 \to 0 \ldots$$

and

$$\ldots 0 \to 0 \to C_n(K) \xrightarrow{\partial} C_{n-1}(K) \to \cdots \to C_1(K) \xrightarrow{\partial} C_0(K) \xrightarrow{\varepsilon} \mathbb{Z} \to 0 \ldots$$

are called the *chain complex* of K and the *augmented chain complex* of K, respectively. The main property of the boundary homomorphism is the following.

Proposition 4.6. *For any p, the homomorphism*

$$\partial \circ \partial = \partial_{p-1} \circ \partial_p : C_p(K) \to C_{p-2}(K)$$

is trivial. Also $\varepsilon \circ \partial_1 : C_1(K) \to \mathbb{Z}$ is trivial.

Proof The last statement is straightforward to check; for the first one it is enough to prove that, for any p-simplex $\sigma = (v^0 \ldots v^p)$ of K, with $p \geqslant 2$,

$$\partial_{p-1}(\partial_p \sigma) = 0.$$

Now

$$\partial_{p-1}(\partial_p \sigma) = \sum_{i=0}^{p} (-1)^i \partial_{p-1}(v^0 \ldots \widehat{v^i} \ldots v^p).$$

This is a sum of $p(p+1)$ terms in which each $(p-2)$-dimensional face of σ occurs exactly twice. In fact, consider the $(p-2)$-dimensional face $(v^0 \ldots \widehat{v^i} \ldots \widehat{v^j} \ldots v^p)$, where $i < j$. This occurs once in

$$(-1)^i \partial_{p-1}(v^0 \ldots \widehat{v^i} \ldots v^p), \text{ with sign } (-1)^i(-1)^{j-1}$$

and once in

$$(-1)^j \partial_{p-1}(v^0 \ldots \widehat{v^j} \ldots v^p), \text{ with sign } (-1)^j(-1)^i.$$

The total coefficient of this $(p-2)$-simplex in $\partial_{p-1}(\partial_p \sigma)$ is therefore $(-1)^i(-1)^{j-1} + (-1)^j(-1)^i = 0$. □

Corollary 4.7. *For any p,*

$$\text{Im } \partial_{p+1} \subset \text{Ker } \partial_p.$$

Also
$$\text{Im } \partial_1 \subset \text{Ker } \varepsilon. \qquad \Box$$

Homology groups

Definitions 4.8. Consider the sequence

$$C_{p+1}(K) \xrightarrow{\partial_{p+1}} C_p(K) \xrightarrow{\partial_p} C_{p-1}(K).$$

Ker ∂_p is denoted by Z_p, or $Z_p(K)$, and elements of Z_p are called *p-cycles* (compare 1.19). Z_p is a free finitely generated (f.g.) abelian group. (See A.33)

Im ∂_{p+1} is denoted by B_p, or $B_p(K)$, and elements of B_p are called *p-boundaries* (we also say that these elements *bound*). B_p is a free f.g. abelian group.

Thus the first statement of 4.7 can be summed up by the motto: 'every boundary is a cycle'. The converse, 'every cycle bounds' is false in general – see the discussion preceding 4.4.

Since $B_p \subset Z_p$ there is a quotient group $H_p(K) = Z_p(K)/B_p(K)$, called the *p*th *homology group* of K. It measures the extent to which K has non-bounding *p*-cycles i.e. the extent to which K has '*p*-dimensional holes' – compare the Introduction, where some examples with $p = 1$ were given. Note that $H_p(K)$ is a finitely generated abelian group.

The quotient group Ker $\varepsilon/B_0(K)$ is denoted by $\widetilde{H}_0(K)$ and is called the *reduced* zeroth *homology group* of K.

Homology groups 105

4.9. Further terminology and remarks (1) Let K^r denote the r-skeleton of K – that is the subcomplex of K consisting of all simplexes of dimension $\leqslant r$. Then clearly $C_p(K) = C_p(K^r)$ and $Z_p(K) = Z_p(K^r)$ provided $r \geqslant p$, and $H_p(K) = H_p(K^r)$ provided $r \geqslant p+1$.

(2) Taking $r = 1$ in (1) we have $Z_1(K) = Z_1(K^1)$. Since K^1 is an oriented graph we can use the method of Chapter 1 to write down a basis for $Z_1(K^1)$. Explicitly, when K is connected, a basis for $Z_1(K) = Z_1(K^1)$ is obtained as follows (compare 1.20): choose a maximal tree T in K' and let $e^1, \ldots, e^\mu$ be the edges of K not in T. Let z^i ($i = 1, \ldots, \mu$) be the 1-cycle associated to the unique loop on $T + e^i$, the coefficient of e^i in z^i being +1. Then $z^1, \ldots, z^\mu$ is the required basis. If K is not connected, then a basis is given by finding a basis for the cycles in each component.

Notice that, according to 1.23, $\tilde{H}_0(K) = 0$ when K is connected.

(3) If K is n-dimensional, then $C_{n+1}(K) = 0$ so that $B_n(K) = 0$ and $H_n(K) = Z_n(K)/0 \cong Z_n(K)$. Perhaps it is being too pedantic to distinguish between Z_n and $Z_n/0$, so let us write $H_n(K) = Z_n(K)$. Note that Z_n is free abelian since it is a subgroup of C_n (see A.33): thus *top dimensional homology groups are free abelian*.

In the case $n = 1$, i.e. K is an oriented graph, we have $H_1(K) = Z_1(K)$ so that, if K is connected, $H_1(K) \cong \mathbb{Z}^{\mu(K)}$ and free generators of $H_1(K)$ are given by the basic cycles described in (2) above. These are the 'independent loops' or 'independent holes' in the graph.

For any K of dimension n, we have $H_p(K) = 0$ for $p < 0$ and for $p > n$.

(4) An element of $H_p(K)$ is a coset $\bar{z} = z + B_p(K)$ where $z \in Z_p(K)$. We shall use the notation $\{z\}$ for this coset; it is called the *homology class* of the cycle z. (If several simplicial complexes are under discussion at the same time, the notation $\{z\}_K$ will be used.) Any cycle z' in $\{z\}$ – that is, any cycle $z' = z + b$ for $b \in B_p(K)$ – is called a *representative cycle* for $\{z\}$. We say that z and z' are *homologous* if $z' - z \in B_p(K)$, and write $z \sim z'$; in particular if $z \in B_p(K)$ (i.e. if z bounds) we write $z \sim 0$: z is homologous to zero. Thus $z \sim z' \Leftrightarrow \{z\} = \{z'\}$. An element of $C_p(K)$ which is not a cycle does not represent any homology class. Nevertheless we could (but won't) consider two chains to be homologous if their difference bounds. Note that a chain which bounds (is homologous to zero) is necessarily a cycle, by 4.7.

(5) As in the case of graphs (see 1.21(3)) the particular orientation given to a simplicial complex does not affect the homology groups up to isomorphism. The orientation is introduced in order to define the boundary homomorphism; nevertheless a theory does exist for unoriented simplicial complexes and this

theory is explained in Chapter 8. For simplicial complexes of dimension > 1 the unoriented theory is a weaker tool than the oriented theory.

Zeroth homology groups Let K be an oriented simplicial complex; the object is to calculate $H_0(K)$. A vertex v of K can also be regarded as a 0-cycle (with the coefficient of v equal to 1 and other coefficients zero), and there is a simple geometrical interpretation of homology between vertices:

Proposition 4.10. $v \sim v' \iff v$ and v' lie in the same component of K. (See 3.17(4))

Proof If v and v' lie in the same component of K, then a path from v' to v will have an associated 1-chain (1.18) c with the property $\partial c = v - v'$; consequently $v \sim v'$. For the converse, we have to use $v \sim v'$ to construct a path from v' to v. The reader can probably convince himself that this is possible by drawing a few pictures: alternatively here is a sketch of a proof by induction. Suppose that $v - v' = \partial c$, and let us write, temporarily, $|c|$ for the sum of the moduli of the coefficients in c; the induction is on $|c|$ and the case $|c| = 1$ is left to the reader. Suppose, to simplify notation, that all 1-simplexes of K with v as an end-point are oriented towards v. Then c must contain at least one such 1-simplex, e say, with positive coefficient. Writing $e = (v''v)$, we have $|c - e| < |c|$ and $\partial(c - e) = v'' - v'$ so that by induction there is a path from v' to v'', and hence via e to v. $\square$

Now let K be a non-empty oriented simplicial complex with say k components $K_1, \ldots, K_k$, and let v^i be a vertex in K_i for $i = 1, \ldots, k$.

Proposition 4.11. $H_0(K)$ is freely generated by the homology classes $\{v^i\}$ for $i = 1, \ldots, k$. Thus $H_0(K) \cong \mathbb{Z}^k$ (also $H_0(\emptyset) = 0$ since $C_p(\emptyset) = 0$ for all p).

Proof The implication '$\Longleftarrow$' of 4.10 shows that $\{v^1\}, \ldots, \{v^k\}$ generate $H_0(K) = C_0(K)/B_0(K)$ (note that $Z_0(K) = C_0(K)$). To show that they are free generators suppose $\lambda_1 v^1 + \cdots + \lambda_k v^k = \partial c$ for some $c \in C_1(K)$. Write $c = c^1 + \cdots + c^k$ where c^i contains only 1-simplexes in K_i ($i = 1, \ldots, k$). Then ∂c^i contains only vertices in K_i, so that we deduce $\lambda_i v^i = \partial c^i$ for $i = 1, \ldots, k$. Taking the augmentation (4.5) of both sides, and using (4.6) $\varepsilon \circ \partial = 0$ we find $\lambda_i = 0$ for $i = 1, \ldots, k$. $\square$

This shows that $H_0(K)$ measures a very fundamental geometrical property of K, namely the number of separate pieces (components) it has.

Remark 4.12. In fact it is always true that, in the above notation, $H_p(K) \cong H_p(K_1) \oplus \cdots \oplus H_p(K_k)$ for all p. This is a special case of a theorem (7.2) to be proved later on homology groups of unions of not necessarily disjoint simplicial complexes. It is also not hard to give a direct proof (start with $k = 2$).

Now consider the sequence

$$C_1(K) \xrightarrow{\partial} C_0(K) \xrightarrow{\varepsilon} \mathbb{Z}.$$

We know that Im $\partial \subset$ Ker $\varepsilon \subset C_0(K)$ by 4.7, and applying A.16, Ex(7), this gives a short exact sequence

$$0 \to \frac{\text{Ker}\,\varepsilon}{\text{Im}\,\partial} \to \frac{C_0}{\text{Im}\,\partial} \to \frac{C_0}{\text{Ker}\,\varepsilon} \to 0$$

i.e.

$$0 \to \widetilde{H}_0(K) \to H_0(K) \to \frac{C_0}{\text{Ker}\,\varepsilon} \to 0.$$

Now suppose that K is non-empty; then ε is epic (surjective) and $C_0/\text{Ker}\,\varepsilon \cong$ Im $\varepsilon = \mathbb{Z}$. Hence the sequence is split (A.19), and the following result holds.

Proposition 4.13. *For a non-empty K,*

$$H_0(K) \cong \widetilde{H}_0(K) \oplus \mathbb{Z}.$$

Also $\qquad\qquad H_0(\emptyset) = \widetilde{H}_0(\emptyset) = 0.$ $\square$

Note that this confirms that, for a connected K, $\widetilde{H}_0(K) = 0$ (4.9(2)). Indeed the main reason for introducing $\widetilde{H}_0(K)$ is that it is often 0 and this makes calculations with exact sequences more straightforward. In the notation of 4.11, the classes represented by $v^i - v^1$ for $i = 2, \ldots, k$ freely generate $\widetilde{H}_0(K)$, when $k \geqslant 2$.

First homology groups Given an oriented connected simplicial complex K we now seek to calculate $H_1(K)$. When K is a graph (dim $K = 1$) there is no problem: we just calculate $Z_1(K)$ as in Chapter 1 (see 4.9(2) and (3)). Thus in this case $H_1(K) \cong \mathbb{Z}^{\mu(K)}$ with generators $\{z^1\}, \ldots, \{z^\mu\}$.

In general there is a nontrivial subgroup $B_1(K)$ of $Z_1(K)$ to factor out:

$$C_2(K) \xrightarrow{\partial_2} C_1(K) \xrightarrow{\partial_1} C_0(K)$$

$$H_1(K) = \text{Ker}\,\partial_1/\text{Im}\,\partial_2 = Z_1(K)/B_1(K).$$

To calculate this quotient in a particular instance we use the method of presentations; see the Appendix but note that the examples we actually meet are very simple and require only a minimum of technique. The example in A.25 is a typical calculation. First we find a basis for $Z_1(K) = Z_1(K^1)$ by the method of 4.9(2), then we find generators for $B_1(K)$. The generators we use are $\partial_2 t$ where t runs over the oriented 2-simplexes of K. A presentation of $H_1(K)$ is, when K is connected,

4.14 $\begin{pmatrix} \text{cycles } z^1, \ldots, z^\mu, \mu = \mu(K^1), \text{ associated to} \\ \text{basic loops on } K^1 \text{ (see 4.9(2))} \\ \text{a relation } \partial_2 t = 0 \text{ for each oriented 2-simplex of } K. \end{pmatrix}$

Direct calculations using 4.14 are rather tedious, and we shall not need to make many of them. Here is an example.

4.15. Möbius Band The diagram shows a triangulation K of the Möbius band, and the heavily drawn edges are ones whose removal leaves a maximal tree in K^1. It is a good idea to label these edges $e^1, \ldots, e^\mu$ (in this example $\mu = 6$) and the remaining edges $e^{\mu+1}, \ldots, e^\alpha$ ($\alpha = \alpha_1(K)$).

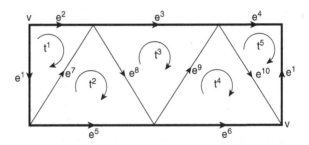

The reason for this is as follows. Suppose that z is any 1-cycle on K; then by 1.20

4.16 $\begin{aligned} z &= \lambda_1 e^1 + \cdots + \lambda_\mu e^\mu + \lambda_{\mu+1} e^{\mu+1} + \cdots + \lambda_\alpha e^\alpha \\ \Rightarrow z &= \lambda_1 z^1 + \cdots + \lambda_\mu z^\mu \end{aligned}$

where z^i ($i = 1, \ldots, \mu$) is the 1-cycle associated with the unique loop on (maximal tree $\cup$ e^i), the 1-cycle being oriented so that it contains e^i with coefficient $+1$. This is of great assistance when expressing $\partial_2 t$ as a linear combination of $z^1, \ldots, z^\mu$.

Let $z^1, \ldots, z^6$ be as above. Thus

$$z^1 = e^1 + e^7 + e^8 + e^9 + e^{10}, \quad z^2 = e^2 + e^8 + e^9 + e^{10}$$

and so on, but there is no need to write them all out. In fact, $B_1(K)$ is generated by the following elements of $Z_1(K)$

$$\partial_2 t^1 = e^2 - e^7 - e^1 = z^2 - z^1 \quad \text{(see 4.16 above)}$$
$$\partial_2 t^2 = e^7 + e^8 - e^5 = -z^5$$
$$\partial_2 t^3 = e^3 - e^9 - e^8 = z^3$$
$$\partial_2 t^4 = e^9 + e^{10} - e^6 = -z^6$$
$$\partial_2 t^5 = e^4 - e^1 - e^{10} = z^4 - z^1.$$

Thus $H_1(K)$ has presentation

$$\left(\begin{matrix} z^1, z^2, z^3, z^4, z^5, z^6 \\ z^2 - z^1 = -z^5 = z^3 = -z^6 = z^4 - z^1 = 0 \end{matrix} \right).$$

(Remember that the coset symbol { } is conventionally omitted when writing down presentations; see A.21.) This shows that all the generators can be expressed in terms of z^1, and (compare A.24), $H_1(K)$ is isomorphic to the abelian group with presentation

$$\left(\begin{matrix} z^1 \\ \text{[no relations]} \end{matrix} \right.$$

which is infinite cyclic. Furthermore it shows that $\{z^1\}_K$ generates $H_1(K)$. Using the relations alternative generators are $\{z^2\}$, or $\{z^4\}$, or $\{z^2 + z^3\}$....

Notice that the cycle z^1 goes 'once round' the Möbius band. Thus the result says that a 1-cycle going 'once round' does not bound anything, and that any 1-cycle is, up to homology, a multiple of this one. For example $e^2 + e^3 + e^4 + e^5 + e^6$ is a 1-cycle going round the rim of the band, and

$$\{e^2 + e^3 + e^4 + e^5 + e^6\} = \{z^2 + z^3 + z^4 + z^5 + z^6\} \quad \text{by 4.16}$$
$$= \{z^2\} + \{z^3\} + \{z^4\} + \{z^5\} + \{z^6\}$$
$$= \{z^1\} + 0 + \{z^1\} + 0 + 0$$

from the presentation of $H_1(K)$

$$= 2\{z^1\}.$$

Indeed it is geometrically clear that the rim is a circle going twice round before joining up. The following picture shows more informally that the rim a is homologous to twice a circle b going round the middle of the band.

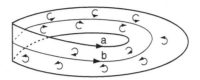

The orientations shown are coherent (2.3) except across the centre line b. Thus $a - 2b$ bounds the whole oriented area (i.e. if the area were broken up into triangles, oriented as shown, the sum of their boundaries would be $a - 2b$), so that $a \sim 2b$.

We know from 4.11 that $H_0(K) \cong \mathbb{Z}$, with generator the homology class of any vertex. To find $H_2(K)$ we need Ker $(\partial_2 : C_2(K) \to C_1(K))$, since $C_3(K) = 0$. Now a general element of $C_2(K)$ is $c = \sum_{i=1}^{5} \lambda_i t^i$; suppose that $\partial_2 c = 0$. The edge e^2 occurs only in $\partial_2 t^1$; hence t^1 must have coefficient zero in c. Similar arguments show that t^2, t^3, t^4, t^5 all have coefficient zero in c, so that $c = 0 : H_2(K) = 0$. An alternative way of doing this is to write down ∂c explicitly, using $\partial t^1 = e^2 - e^7 - e^1$ etc., and to deduce from $\partial c = 0$ that $\lambda_1 = \ldots = \lambda_5 = 0$.

Thus $H_0(K) \cong \mathbb{Z}$ generator $\{v\}$

$H_1(K) \cong \mathbb{Z}$ generator $\{e^1 + e^7 + e^8 + e^9 + e^{10}\}$

$H_2(K) = 0$.

Examples 4.17. (1X) Triangulate the cylinder (3.22(3)) and show that the homology groups are

$H_0 \cong \mathbb{Z}$; $H_1 \cong \mathbb{Z}$ (generator a cycle going round one end of the cylinder)

$H_2 = 0$.

Thus up to isomorphism the groups are the same as for the Möbius band. This fact becomes obvious by the method of collapsing (3.30) applied to homology groups (6.6). See 6.7(1).

(2X) Show that, for any p,

$$C_p(K) \cong Z_p(K) \oplus B_{p-1}(K).$$

[Hint. Form a short exact sequence from $\partial_p : C_p(K) \to C_{p-1}(K)$.]

(3X) Using $B_p(K) \subset Z_p(K) \subset C_p(K)$ and A.16, Ex. (7) show that

$$C_p(K)/B_p(K) \cong H_p(K) \oplus B_{p-1}(K).$$

Homology groups

This shows that the torsion subgroup of $H_p(K)$ (the subgroup of $H_p(K)$ consisting of those elements which are of finite order) is isomorphic to the torsion subgroup of C_p/B_p (since B_{p-1} is free abelian). In particular the torsion subgroup of $H_p(K)$ is determined by the homomorphism $\partial_{p+1} : C_{p+1}(K) \to C_p(K)$.

Calculate the group $C_1(K)/B_1(K)$ when K is an n-sided polygon. Note that the result depends on n, whereas $H_1(K) \cong \mathbb{Z}$ for any n since $\mu(K) = 1$ (see 4.9(3)). For example if K is an n-sided polygon then $C_1(K)/B_1(K)$ is isomorphic to $\mathbb{Z}^n$, either because $B_1(K)$ is trivial or, using the above isomorphism, $H_1(K) \cong \mathbb{Z}$, $B_0(K) \cong \mathbb{Z}^{n-1}$, generated by the boundaries of $n-1$ of the edges.

(4X) Calculate the homology groups of the torus given in 2.2(3), or in 3.22(1). (Of course you must first orient the 1- and 2-simplexes, but it doesn't matter how – see 4.9(5).) This homology calculation is about as complicated as one normally wants to do 'from first principles'. Verify that $H_0 \cong \mathbb{Z}$; $H_1 \cong \mathbb{Z} \oplus \mathbb{Z}$ with generators $\{a+b+c\}$ and $\{d+e+f\}$ in 2.2(3); $H_2 \cong \mathbb{Z}$ with generator the sum of all triangles of the torus, coherently oriented.

(5X) Orient the simplicial complex shown, which is a projective plane (2.7(3)), and verify that the homology groups are $H_0 \cong \mathbb{Z}$; $H_1 \cong \mathbb{Z}_2$ with generator $\{a+b+c\}$; $H_2 = 0$.

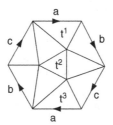

Note that H_1 is a finite group here; in fact $2(a+b+c) = 0$, i.e. $2(a+b+c) \sim 0$. This can be seen directly by adding up the boundaries of the ten triangles in the picture, all oriented clockwise. (The three labelled triangles are for later reference.)

We shall not have to calculate, directly from the definition, many homology groups $H_p(K)$ for $p > 1$. There is one case of interest, however, in connexion with the orientability of closed surfaces.

Proposition 4.18. *Let K be a closed surface (2.1 or 3.29(2)). Then*

$$H_2(K) \cong \begin{cases} 0 & \text{if } K \text{ is non-orientable (2.3)} \\ \mathbb{Z} & \text{if } K \text{ is orientable.} \end{cases}$$

A generator for $H_2(K)$, when K is orientable, is given by the homology class of the sum of all 2-simplexes of K, coherently oriented.

To be precise: K is already oriented, and, apart from this, the 2-simplexes can be coherently oriented. The generator of $H_2(K)$ is $\Sigma(\pm t^i)$, the sum being over all oriented 2-simplexes t^i of K, with sign $+$ or $-$ according as the orientation of t^i as a simplex of K does or does not agree with that in the chosen coherent orientation. Note that $H_2(K) = Z_2(K)$ since K is two-dimensional (see 4.9(3)). A generator of $H_2(K)$, when K is an orientable closed surface, is called a *fundamental cycle*, or *fundamental class* for K. Up to sign it is determined by K.

Proof Suppose first that K is orientable. For simplicity let us suppose that the orientation of K is such that the 2-simplexes receive orientations which are already coherent. We seek $Z_2(K)$ (compare 4.9(3)), so let $z \in Z_2(K)$, and suppose $z \neq 0$ so that some oriented 2-simplex t occurs with coefficient $\lambda \neq 0$ in z. Let t' be any other oriented 2-simplex. Then (see 2.2(5)) there is a sequence of triangles of K connecting t and t', two consecutive members of the sequence always having an edge in common. From this and $\partial z = 0$ it follows that every triangle in the sequence, and in particular t', also has coefficient λ in z. Hence z is a multiple of the sum of all triangles, coherently oriented, and $Z_2(K)$ is infinite cyclic with this sum as generator.

Now suppose that $H_2(K) \neq 0$ and let z be a non-zero element of $Z_2(K)$. Exactly as above we can deduce that the coefficient of each oriented 2-simplex in z must be, up to sign, the same non-zero integer (only up to sign because the triangles of K are no longer assumed coherently oriented). Now define a new orientation of the triangles of K by changing the orientation of those with negative coefficient in z. It is easy to see (using $\partial z = 0$) that this coherently orients the triangles of K: K is orientable. $\square$

Thus H_2 distinguishes algebraically between orientable and non-orientable surfaces. Fundamental cycles for the tori drawn in 2.2(3) and 3.22(1) are obtained by orienting all the triangles in the pictures clockwise, or alternatively all anti-clockwise. Of course this doesn't work for the Klein bottle or projective plane of 2.2(4), for such an orientation will not be coherent across, for example, the edge d.

Relative homology groups

In Chapter 9 we shall investigate some of the properties of a graph which is realized as a subcomplex of a surface. This is a situation involving a pair of simplicial complexes – or, as we shall say, a relative situation – and calls for algebraic tools which involve both complexes at once. In particular consider the graph in a torus drawn below.

Relative homology groups

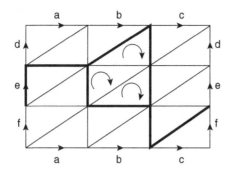

There is a region bounded by five edges and containing the three oriented triangles. The sum of the boundaries of these three triangles is certainly not zero, but it does contain only edges in the graph, and the sum of the three triangles is called a relative cycle. Thus relative cycles have a connexion with regions, and this is exploited in Chapter 9.

Definition 4.19. Let K be an oriented simplicial complex of dimension n and L be a subcomplex of K (L is therefore oriented in a natural way by K). For any p, $0 \leqslant p \leqslant n$, the pth *chain group of K modulo L*, or the pth *relative chain group* (of the pair (K, L)) is the subgroup of $C_p(K)$ consisting of those p-chains on K in which the coefficient of every simplex of L is zero. It is denoted $C_p(K, L)$, and an element of $C_p(K, L)$ is called a *p-chain of K modulo L*, or a *relative p-chain*. (There is always a choice of terminology between 'relative' and 'modulo'. In the definitions below we shall usually give the former and leave the reader to give expression to the latter. Also 'modulo', when it occurs, will always be abbreviated to 'mod'.) For $p > n$ or $p < 0$, $C_p(K, L)$ is defined to be zero.

Note that $C_p(K, \emptyset) = C_p(K)$. The following definition of 'relative boundary' also reduces to the old definition of boundary when $L = \emptyset$.

Definition 4.20. With the notation of 4.19, we define

$$j = j_q \; : \; C_q(K) \to C_q(K, L)$$

to be the homomorphism, which changes to zero (or leaves at zero) the coefficient of every simplex in L (for $q < 0$ or $q > n$, $j_q = 0$). Now define

$$\widehat{\partial} = \widehat{\partial}_p \; : \; C_p(K, L) \to C_{p-1}(K, L)$$

by $\widehat{\partial} c = j_{p-1}(\partial_p c) \quad \text{for all} \quad c \in C_p(K, L)$

$\widehat{\partial}$ called the *relative boundary homomorphism*. Thus if i_p denotes the inclusion of $C_p(K, L)$ in $C_p(K)$, then $\widehat{\partial}_p = j_{p-1} \circ \partial_p \circ i_p$. We may say that j 'forgets L'

and then the relative boundary of a chain is the ordinary boundary followed by forgetting L.

Proposition 4.21. *For any p, the homomorphism*

$$\widehat{\partial} \circ \widehat{\partial} = \widehat{\partial}_{p-1} \circ \widehat{\partial}_p \; : \; C_p(K,L) \to C_{p-2}(K,L)$$

is trivial. □

The proof is similar to that of 4.6 and will be omitted.

The remaining definitions are precisely analogous to those for the 'absolute' case $L = \emptyset$. They are summarized below.

Definitions 4.22. *Chain complex* of (K,L):

$$\ldots 0 \to C_n(K,L) \xrightarrow{\widehat{\partial}} C_{n-1}(K,L) \xrightarrow{\widehat{\partial}} \ldots$$
$$\xrightarrow{\widehat{\partial}} C_1(K,L) \xrightarrow{\widehat{\partial}} C_0(K,L) \to 0.$$

(There is no augmented relative chain complex.)

Group of *relative p-cycles* $= Z_p(K,L) = \text{Ker } \widehat{\partial}_p$.
Group of *relative p-boundaries* $= B_p(K,L) = \text{Im } \widehat{\partial}_{p+1}$.

By 4.21, $B_p(K,L) \subset Z_p(K,L)$; the quotient group

$$H_p(K,L) = Z_p(K,L)/B_p(K,L)$$

is the pth *relative homology group* of the pair (K,L). All these reduce to the previous definitions when $L = \emptyset$.

If $z \in Z_p(K,L)$ then $z + B_p(K,L) \in H_p(K,L)$ is denoted by $\{z\}$ or $\{z\}_{K,L}$ and is called the *relative homology class* of z. Any element of $\{z\}$, i.e. any $z + b$ for $b \in B_p(K,L)$, is called a *representative relative cycle* for $\{z\}$. Two relative cycles z and z' are said to be *homologous mod L* (written $z \stackrel{L}{\sim} z'$) if $\{z\} = \{z'\}$.

Remarks and Examples 4.23. (1) If $K = L$ then all the groups and homomorphisms are trivial. If $\dim K = n$, then $H_p(K,L) = 0$ for $p < 0$ or $p > n$, and $H_n(K,L) = Z_n(K,L)$ is free abelian.

(2) $H_p(K,L)$ is not in general a subgroup of $H_p(K)$. For example let K be a 1-simplex $e = (vw)$ together with its end-points and let L be the end-points. Then $H_1(K) = 0$ since K is a tree (see the preamble to 4.14), but $H_1(K,L) \cong \mathbb{Z}$,

with generator $\{e\}$. For $C_0(K,L) = 0$, so $\widehat{\partial}_1 = 0$ and $Z_1(K,L) = C_1(K,L)$ is infinite cyclic with generator e.

(3) More generally than (2), let K be any oriented graph and $L = K^0$, the set of vertices of K. Then $C_0(K,L) = 0$, $\widehat{\partial}_1 = 0$ and $Z_1(K,L) = C_1(K,L)$ is a free abelian group with one generator for each edge of K.

If K is any oriented connected graph and L is a maximal tree (1.10) for K, then L contains all the vertices of K and so $C_0(K,L) = 0$, $\widehat{\partial}_1 = 0$ and $Z_1(K,L) = C_1K,L)$ is a free abelian group with one generator for each edge of K not in L – hence of rank $\mu(K)$.

(4) Let K^r denote the r-skeleton of K. Then the chain complex of (K^r, K^{r-1}) is

$$\ldots \to 0 \to C_r(K^r, K^{r-1}) \to 0 \to \ldots$$

with only one non-zero group. Thus $H_p(K^r, K^{r-1})$ is zero when $p \neq r$, and is free abelian of rank $\alpha_r(K)$ (the number of r-simplexes of K) when $p = r$. As an example, let K be a 2-simplex and all its faces, and $r = 2$, so that $K^r = K$, K^{r-1} = the set of edges and vertices of K. Then $H_2(K, K^1)$ is free abelian of rank 1, generated by the homology class of the 2-simplex. In this case $H_1(K, K^1) = 0 = H_0(K, K^1)$.

The next proposition enables relative cycles and boundaries to be recognized as such without undue difficulty. It is convenient, now and later, to make the following convention. When L is a subcomplex of K, we regard $C_p(L)$ as a subgroup of $C_p(K)$, namely the subgroup consisting of those p-chains on K in which the coefficient of every p-simplex not in L is zero. (Thus, with this convention, $C_p(K)$ is the internal direct sum of $C_p(L)$ and $C_p(K,L)$.) In order that the convention shall not lead us into error, it is essential that the boundary of a p-chain on L should be the same whether we regard it in its own right or according to the convention as a special kind of p-chain on K. That this is indeed the case follows from the fact that L is a subcomplex of K and that therefore if $\sigma \in L$ then all the faces of σ also belong to L.

Proposition 4.24. *(1) Let, in the usual notation, $c \in C_p(K,L)$. Then c is a relative cycle, i.e. belongs to $Z_p(K,L)$, if and only if $\partial c \in C_{p-1}(L)$.*

(2) Let $c \in C_p(K,L)$. Then c is a relative boundary, i.e. $c \in B_p(K,L)$, if and only if there is an (absolute) p-chain $c' \in C_{p+1}(K)$ such that $\partial c' - c \in C_p(L)$.

Proof (1) is virtually immediate from the definitions (and the above convention), so the proof will be omitted. (2) can be a very confusing statement to verify, so here is a full proof.

Suppose that $c \in B_p(K, L)$, i.e. $c = \widehat{\partial} c'$ for some $c' \in C_{p+1}(K, L)$. We show that this c' (which does of course belong to $C_{p+1}(K)$) satisfies $\partial c' - c \in C_p(L)$. In fact
$$jc = c = \widehat{\partial} c' = j \partial c', \quad \text{so that} \quad j(c - \partial c') = 0.$$
From the definition of j, this implies $\partial c' - c \in C_p(L)$.

Suppose conversely that c', as given in (2), exists. Then let $c'' = jc' \in C_{p+1}(K, L)$. We show $\widehat{\partial} c'' = c$, which proves that $c \in B_p(K, L)$.

We shall use the following simple but useful trick. Let $\rho : C_{p+1}(K) \to C_{p+1}(L)$ denote the 'restriction' homomorphism which puts at zero the coefficient of any simplex not in L. Then clearly, for any $c' \in C_{p+1}(K)$, $c' = jc' + \rho c'$.

We are given $j(\partial c' - c) = 0$; also $jc = c$ since $c \in C_p(K, L)$. Hence

$$c = jc = j\partial c' = j\partial jc' + j\partial \rho c' = j\partial jc' \quad \text{since} \ \partial \rho c' \in C_p(L)$$
$$= \widehat{\partial} c'' \quad \text{by definition.} \qquad \square$$

Zeroth relative homology groups As an example of 4.24, let K be connected and L non-empty. We show $H_0(K, L) = 0$, i.e. $B_0(K, L) = C_0(K, L)$. Let v be a vertex of K not in L (if no such vertex exists then the result is true since $C_0(K, L) = 0$), and w a vertex of L. Join w to v by a path in K; the associated 1-chain (1.18) c' has $\partial c' = v - w$, so that $\partial c' - v \in C_0(L)$, i.e. (by 4.24) $v \in B_0(K, L)$. Hence every vertex of K not in L is a relative boundary, which implies the result.

The general result on $H_0(K, L)$ is as follows. Details of the proof are left to the reader.

Proposition 4.25. *Let K be an oriented simplicial complex and L a subcomplex of K. Then $H_0(K, L) = 0$ provided L has at least one vertex in every component of K, but $H_0(K, L)$ acquires an infinite cyclic summand for every component of K not containing any vertex of L. Free generators for $H_0(K, L)$ are the relative homology classes of one vertex from each of these latter components.* $\qquad \square$

Remark 4.26. (compare 4.12) Let K have components $K_1, \ldots, K_k$ and write $L_1 = L \cap K_1, \ldots, L_k = L \cap K_k$. Then it can be shown that, for all p,

$$H_p(K, L) \cong H_p(K_1, L_1) \oplus \cdots \oplus H_p(K_k, L_k).$$

First relative homology groups To calculate $H_1(K,L)$ we need a basis for $Z_1(K,L)$ and generators for $B_1(K,L)$. The method of presentations will then give the result as in the absolute ease. The easiest case is when all the vertices of K belong to L, for then $C_0(K,L) = 0$ so that $Z_1(K,L) = C_1(K,L)$ and a basis is given by the edges of K not in L. Generators for $B_1(K,L)$ are simply $\widehat{\partial_2}t$ where t runs through the oriented triangles of K not in L.

For completeness, here is a method for writing down a basis of $Z_1(K,L)$ in general, followed by a very brief sketch of the proof. We assume K is connected (compare 4.26) to simplify the statement.

Let K be a connected oriented simplicial complex and L a subcomplex of K, with components $L_1, \ldots, L_\ell$, say. Let T_α be a maximal tree in the 1-skeleton of L_α ($\alpha = 1, \ldots, \ell$) and let T be a maximal tree in the 1-skeleton of K which contains $T_1, \ldots, T_\ell$ (compare 1.12(1)). For each edge e^i of K which is in neither T nor L let z^i be the 1-cycle on K given by the unique loop on $T + e^i$ such that the coefficient of e^i in z^i is $+1$. Thus the number of z^i is

$$m = \mu(K) - \mu(L_1) - \cdots - \mu(L_\ell).$$

Next let $v^1, \ldots, v^\ell$ be vertices of K in, respectively, the components $L_1, \ldots, L_\ell$ of L, and let c^α ($\alpha = 2, \ldots, \ell$) be the 1-chain associated to some path in K from v^1 to v^α. (Thus if $\ell = 0$ or $\ell = 1$ there is no c^α to consider.)

Proposition 4.27. ★ *The following elements of $Z_1(K,L)$ are free generators:*

$$jz^1, \ldots, jz^m, jc^2, \ldots, jc^\ell.$$

Sketch of Proof The proof depends on the following sequence.

$$0 \to Z_1(L) \xrightarrow{i} Z_1(K) \xrightarrow{j} Z_1(K,L) \xrightarrow{\partial_*} H_0(L) \xrightarrow{i_*} H_0(K) \to 0$$

where i = inclusion, j is the j of 4.20 restricted to $Z_1(K)$, $\partial_* z = \{\partial z\}_L$, $i_*\{w\}_L = \{w\}_K$. The sequence is a special case of the 'homology sequence of a pair' (see 6.3(6)) and is exact, provided L is non-empty. The case $L = \emptyset$ ($\ell = 0$) is covered by 1.20. The proof of exactness will be left to the reader – or he can wait for the general theorem. From the exact sequence we deduce two short exact sequences:

$$0 \to \operatorname{Im} j \xrightarrow{i} Z_1(K,L) \xrightarrow{\partial_*} \operatorname{Im} \partial_* \to 0 \qquad (*)$$

$$\text{and } 0 \to Z_1(L) \xrightarrow{i} Z_1(K) \xrightarrow{j} \operatorname{Im} j \to 0 \qquad (**)$$

(each i being an inclusion).

118   Homology groups

Free generators for $Z_1(L)$ are obtained by putting back, one by one, the edges of L_1 not in T_1, then the edges of L_2 not in T_2, etc. Free generators for $Z_1(K)$ are all of these, plus $z^1, \ldots, z^m$. It follows from (**) that free generators for Im j are $jz^1, \ldots, jz^m$.

Free generators for Im $\partial_* =$ Ker i_* are $\{v^2 - v^1\}, \{v^3 - v^1\}, \ldots, \{v^\ell - v^1\}$, since $i_*(\lambda_1\{v^1\}_L + \cdots + \lambda_\ell\{v^\ell\}_L) = (\lambda_1 + \cdots + \lambda_\ell)\{v^1\}_K$. Finally, $\partial_* jc^\alpha = \partial_*(c^\alpha - \rho c^\alpha)$ (where ρ is the 'restriction' which puts at zero the coefficients of simplexes not in L) $= \{\partial c^\alpha\}_L - \{\partial \rho c^\alpha\}_L = \{v^\alpha - v^1\}_L - 0$, since $\rho c^\alpha \in C_1(L)$ so that $\{\partial \rho c^\alpha\}_L = 0$. Thus $jc^2, \ldots, jc^\ell$ are elements of $Z_1(K, L)$ which go, under ∂_*, to the free generators of Im ∂_*. The result follows from (*) and the last sentence of A.20(3). ★ □

Remark 4.28. ★ The following method for expressing a given $z \in Z_1(K, L)$ as an integer linear combination of the free generators listed in 4.27 is a straightforward consequence of the above proof.

Coefficient of jz^i = coefficient of e^i in z ($i = 1, \ldots, m$).

Coefficient of jc^α = coefficient of $\{v^\alpha\}$ in $\partial_* z$ ($\alpha = 2, \ldots, \ell$)

= the sum of the coefficients of those vertices of

L^α in $\partial z \in C_0(L)$. ★

Examples 4.29. (1) Let K be the cylinder drawn below and L the (heavily drawn) subcomplex consisting of the two ends.

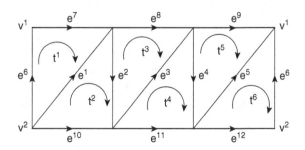

In this case L contains all the vertices of K, so that $C_0(K, L) = 0$, and $Z_1(K, L) = C_1(K, L)$ with basis $e^1, \ldots, e^6$. Generators for $B_1(K, L)$ are $\widehat{\partial} t^1, \ldots, \widehat{\partial} t^6$ and a presentation for $H_1(K,L)$ is

$$\begin{pmatrix} e^1, \ldots, e^6 \\ -e^1 + e^6 = e^1 + e^2 = -e^2 - e^3 = e^3 + e^4 = -e^4 - e^5 = e^5 - e^6 = 0 \end{pmatrix}$$

Thus $H_1(K,L)$ is infinite cyclic, with generator $\{e^6\}$ for example.

★ If we use 4.27 to obtain a basis for $Z_1(K,L)$, we have $\ell = 2$ and taking L_1 to be the top of the cylinder and L_2 the bottom, we can take $T_1 = \{e^7, e^8, v^1, v^3, v^4\}, T_2 = \{e^{10}, e^{11}, v^2, v^5, v^6\}$ and $T = T_1 \cup T_2 \cup \{e^6\}$. Then $m = 5$ and the cycles $z^1, \ldots, z^5$ from z are obtained from T by replacing, respectively say, the edges $e^1, \ldots, e^5$. The first five generators of $Z_1(K,L)$ are then $e^1 - e^6, e^2 + e^6, e^3 - e^6, e^4 + e^6, e^5 - e^6$. The remaining generator is, say, e^2, which is jc^2 where $c^2 = e^7 + e^2 - e^{10}$ is associated to a path from $v^1 \in L_1$ to $v^2 \in L_2$. To express say $\widehat{\partial} t^4 = e^3 + e^4$ in terms of these we use 4.28: the coefficients of jz^3 and jz^4 are both 1, and the coefficient of jc^2 is 0 since $\partial(e^3 + e^4) = v^6 - v^5$, with sum of coefficients 0.

I leave it to the reader to verify that $H_2(K,L) \cong \mathbb{Z}$ with generator $\{t^1 + \cdots + t^6\}$. Of course $H_0(K,L) = 0$ (4.25).★

★(2X) Take K as in (1), but L this time to be one end, say the top, of the cylinder. Here L does not contain all the vertices of K, so some technique, e.g that of 4.27, is really needed to write down generators for $Z_1(K,L)$. Verify that $H_1(K,L)$ and $H_2(K,L)$ are both trivial.★

(3X) Take K to be the Möbius band given in 4.15 and L to be the rim of the band consisting of the edges $e^2, \ldots, e^6$ and their end-points. Verify that $H_1(K,L) \cong \mathbb{Z}_2$, with generator $\{e^1\}$. Notice that $e^1 + e^1 \overset{L}{\sim} 0$; to see this geometrically refer to the left-hand diagram below.

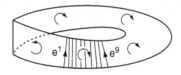 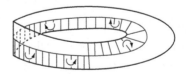

Now $e^1 \overset{L}{\sim} e^9$ since the boundary of the shaded area (oriented as shown) is, apart from pieces of L, $e^1 - e^9$. But also $e^1 \overset{L}{\sim} -e^9$ since the boundary of the rest of the band, oriented as shown, is $e^1 + e^9$ apart from pieces of L. Hence $e^1 \overset{L}{\sim} -e^1$.

Verify that $e^1 + e^7 + e^8 + e^9 + e^{10}$, which goes 'once round' the band is homologous mod L to e^1. This can be seen geometrically by considering the boundary of the shaded region in the right-hand diagram.

(4X) Consider the homomorphism $j : C_1(K) \to C_1(K,L)$. If $z \in Z_1(K)$ then $jz \in Z_1(K,L)$, for

$$\widehat{\partial} jz = j\partial jz = j\partial(z - \rho z) \quad (\rho : C_1(K) \to C_1(L) \text{ puts coefficients of}$$
$$\text{simplexes not in } L \text{ equal to } 0)$$
$$= 0 - j\partial\rho z \text{ since } \partial z = 0$$
$$= 0 \text{ since } \partial\rho z = C_0(L).$$

Thus there is an induced homomorphism, which we still call j,

$$j : Z_1(K) \to Z_1(K,L)$$

(this appeared in the proof of 4.27). Suppose that L is connected (it does not matter whether K is connected). Show that this j is epic (surjective), i.e. that every relative 1-cycle can be 'extended', by adding edges of L, to an absolute 1-cycle. This is plausible from the picture, where the relative cycle w can be extended to an absolute cycle by adding the part of L between the endpoints v and v' of w.

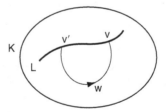

[Hint for the formal proof: let $\varepsilon_K : C_0(K) \to \mathbb{Z}$ and $\varepsilon_L : C_0(L) \to \mathbb{Z}$ be the augmentations (4.5). Show that $\varepsilon_L \partial w = \varepsilon_K \partial w = 0$ (see 4.6) and then use 1.23, which says that $C_1(L) \xrightarrow{\partial} C_0(L) \xrightarrow{\varepsilon_L} \mathbb{Z}$ is exact.] The result is false when L is not connected. For example the relative cycle e^1 on the cylinder in (1) above cannot be extended to an absolute cycle by adding edges in L.

(5X) Take K to be the torus given in 2.2(3), and L to be some particular subcomplex of K. Verify that $H_2(K,L)$ is free of rank equal to the number of regions into which L divides K (in the obvious intuitive sense of 'region'). Note that when L is the 1-skeleton of K this is a special case of 4.23(4), for the regions are then the 2-simplexes.

This will not work in general for a non-orientable surface such as the Klein bottle. For example take $L = \emptyset$; then there is one region but $H_2(K,L) = H_2(K) = 0$ by 4.18. More of this in Chapter 9.

Three homomorphisms

We shall define three homomorphisms

$$i_* : H_p(L) \to H_p(K)$$
$$j_* : H_p(K) \to H_p(K,L)$$
$$\partial_* : H_p(K,L) \to H_{p-1}(L)$$

which, in the next chapter, will be fitted together to form an exact sequence. An alternative approach is sketched in an Appendix to this chapter. As usual K will be an oriented simplicial complex and L a subcomplex of K.

4.30. The homomorphism i_*

Define $i_* : H_p(L) \to H_p(K)$

by $i_*\{z\}_L = \{z\}_K \quad (z \in Z_p(L))$

i.e. $i_*(z + B_p(L)) = z + B_p(K)$.

Thus we simply regard $z \in Z_p(L)$ as a cycle on K and take its homology class there. Of course $\{z\}_K$ can be zero when $\{z\}_L$ is non-zero. For example let K be a 2-simplex and all its faces, and L the boundary of the 2-simplex (the set of proper faces). Then $\{e_1 + e_2 + e_3\}_L$ generates $H_1(K) \cong \mathbb{Z}$, but $\{e^1 + e^2 + e^3\}_K = 0$ since $\partial t = e^1 + e^2 + e^3$.

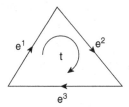

Several things must be checked before the definition of i_* is acceptable. These are

(1) If $z \in Z_p(L)$ then $z \in Z_p(K)$ (recall the preamble to 4.24).
(2) If $z \in B_p(L)$ then $z \in B_p(K)$.
(3) $\{z + z'\}_K = \{z\}_K + \{z'\}_K$ where $z, z' \in Z_p(L)$.

(1) and (2) say that i_* is well-defined, and (3) say that i_* is a homomorphism. All these assertions are straightforward to check.

Examples 4.31. (1) Let K be the Möbius band given in 4.15 and let L be the rim of the band. Then $H_1(L) \cong \mathbb{Z}$, with generator $\{e^2 + e^3 + e^4 + e^5 + e^6\}_L$, and $H_1(K) \cong \mathbb{Z}$, with generator $\{z^1\}_K = \{e^1 + e^7 + e^8 + e^9 + e^{10}\}_K$ (see 4.15). The homomorphism i_* is determined by its effect on the generator of $H_1(L)$, and according to the calculation given in 4.15,

$$i_*\{e^2 + e^3 + e^4 + e^5 + e^6\}_L = \{e^2 + e^3 + e^4 + e^5 + e^6\}_K$$
$$= 2\{z^1\}_K.$$

Thus i_* takes a generator to twice a generator: it is 'multiplication by two'.

(2X) Let K be the cylinder given in 4.29(1), (2) and let L be (i) both ends; (ii) one end of the cylinder. Show that i_* takes (i) each of the generators $\{e^7 + e^8 + e^9\}_L$ and $\{e^{10} + e^{11} + e^{12}\}_L$ of $H_1(L) \cong \mathbb{Z} \oplus \mathbb{Z}$ to a generator of $H_1(K)$; (ii) the generator $\{e^7 + e^8 + e^9\}_L$ of $H_1(L) \cong \mathbb{Z}$ to a generator of $H_1(K)$. What is Ker i_* in case (i)?

(3X) Suppose that K has components $K_1, \ldots, K_k$ and that in addition $L \cap K_1, \ldots, L \cap K_k$ are connected (possibly empty). Show that $i_* : H_0(L) \to H_0(K)$ is a monomorphism (injective). When is it an isomorphism? (In fact it is an isomorphism precisely when $H_0(K, L) = 0$; in particular when K and L are both connected and non-empty.)

(4X) Let K be the oriented simplicial complex obtained from the projective plane drawn in 4.17(5) by removing the three 2-simplexes t^1, t^2, t^3 (and then orienting the remaining simplexes). Let L be the subcomplex of K consisting of the eight 1-simplexes and five 0-simplexes on the boundaries of these three 2-simplexes. Find four 1-cycles z^1, z^2, z^3, z^4 which freely generate $Z_1(L)$. Calculate $H_1(K)$, including generators, and express in terms of them the four homology classes $\{z^\alpha\}_K, \alpha = 1, 2, 3, 4$. Hence find $i_* : H_1(L) \to H_1(K)$; in particular show that it is epic (surjective).

4.32. The homomorphism j_* Define $j_* : H_p(K) \to H_p(K, L)$ by $j_*\{z\}_K = \{jz\}_{K,L}$ for $z \in Z_p(K)$.

This time the following have to be checked:

(1) If $z \in Z_p(K)$ then $jz \in Z_p(K, L)$ (see 4.20 for j).
(2) If $z \in B_p(K)$ then $jz \in B_p(K, L)$.
(3) $\{j(z + z')\}_{K,L} = \{jz\}_{K,L} + \{jz'\}_{K,L}$.

(The proof of (2) is very similar to that of 4.24(2).)

Examples 4.33. (1X) Take K and L as in 4.31(1). Then $H_1(K,L) \cong \mathbb{Z}_2$ with generator $\{e^1\}$ (see 4.29(3)). Thus $j_*\{e^1 + e^7 + e^8 + e^9 + e^{10}\}_K = \{e^1\} + \{e^7\} + \{e^8\} + \{e^9\} + \{e^{10}\}$ (all on $(K,L)) = 5\{e^1\} = \{e^1\}$. Thus j_* takes generator to generator. What is $j_*\{e^6 + e^2 + e^3 + e^4 + e^5\}_K$?

(2X) Take K and L to be as in (i) of 4.31(2). Show that $j_* : H_1(K) \to H_1(K,L)$ is trivial.

4.34. The homomorphism ∂_* Define $\partial_* : H_p(K,L) \to H_{p-1}(L)$

by $\partial_*\{z\}_{K,L} = \{\partial z\}_L \quad z \in Z_p(K,L)$.

This time the following have to be checked:

(1) If $z \in Z_p(K,L)$ then $\partial z \in Z_{p-1}(L)$.
(2) If $z \in B_p(K,L)$ then $\partial z \in B_{p-1}(L)$.
(3) $\{\partial z + \partial z'\}_L = \{\partial z\}_L + \{\partial z'\}_L$.

From 4.24(1), $\partial z \in C_{p-1}(L)$; however $\partial \partial z = 0$ so $\partial z \in Z_{p-1}(L)$. Notice, incidentally, that although ∂z is a boundary, $\{\partial z\}_L$ is not automatically zero. This is because ∂z is not usually the boundary of something in L – though of course it might be, as in the picture of 4.29(4), where z appears as w. For (2), suppose $z = \widehat{\partial} w$ ($w \in C_{p+1}(K,L)$); show that $\partial z = \partial(-\rho \partial w)$, where $\rho : C_p(K) \to C_p(L)$ puts coefficients of simplexes not in L equal to 0. (See the diagram below.) The assertion (3) is immediate.

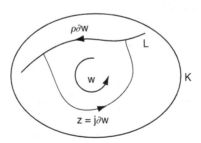

Examples 4.35. (1) Take K, L as in 4.31(1). Then $\partial_* : H_1(K,L) \to H_0(L)$ is trivial – hardly surprising, for any homomorphism $\mathbb{Z}_2 \to \mathbb{Z}$ is trivial.

(2X) Take K, L as in (i) of 4.31(2). Then $\partial_* : H_2(K,L) \to H_1(L)$ takes the generator $\{t^1 + \cdots + t^6\}_{K,L}$ to $\{e^7 + e^8 + e^9\}_L - \{e^{10} + e^{11} + e^{12}\}_L$, while $\partial_* : H_1(K,L) \to H_0(L)$ takes $\{e^6\}_{K,L}$ to $\{v^1\}_L - \{v^2\}_L$.

(3X) The reader may care to verify that the sequence

$$H_2(L) \xrightarrow{i_*} H_2(K) \xrightarrow{j_*} H_2(K,L) \xrightarrow{\partial_*} H_1(L) \xrightarrow{i_*} H_1(K)$$
$$\xrightarrow{j_*} H_1(K,L) \xrightarrow{\partial_*} H_0(L) \xrightarrow{i_*} H_0(K) \xrightarrow{j_*} H_0(K,L) \to 0$$

is exact in each of the cases (1), (2) above. (Compare 6.1.)

(4X) If K and L are both connected, then $\partial_* : H_1(K,L) \to H_0(L)$ is always trivial (compare (1) above).

★ Appendix on chain complexes

This appendix is optional reading, except that in Chapter 5, on the invariance of homology groups, we make brief use of the first two items, 4.36 and 4.37. The main purpose of the appendix is to sketch a systematic approach to the construction of homomorphisms like the three, i_*, j_* and ∂_*, constructed above. An elegant generalization of this approach has been given by D. Puppe (1962).

Definitions 4.36. A *chain complex* (C_p, d_p) is a collection (C_p) of abelian groups, one for each $p \in \mathbb{Z}$, together with a homomorphism $d_p : C_p \to C_{p-1}$ for each p, such that, for all p, $d_{p-1} \circ d_p = 0$. The examples to keep in mind are $C_p = C_p(K)$, $d_p = \partial_p$ where K is an oriented simplicial complex, and $C_p = C_p(K,L)$, $d_p = \widehat{\partial}_p$ where L is a subcomplex of the oriented simplicial complex K. A *chain map* (f_p) from a chain complex (C_p, d_p) to another, (C'_p, d'_p) is a collection of homomorphisms $f_p : C_p \to C'_p$, such that $d'_p \circ f_p = f_{p-1} \circ d_p$ for all p. This means that the diagram

$$\begin{array}{ccccccc}
\cdots \to & C_{p+1} & \xrightarrow{d_{p+1}} & C_p & \xrightarrow{d_p} & C_{p-1} & \to \cdots \\
& \downarrow f_{p+1} & & \downarrow f_p & & \downarrow f_{p-1} & \\
\cdots \to & C'_{p+1} & \xrightarrow{d'_{p+1}} & C'_p & \xrightarrow{d'_p} & C'_{p-1} & \to \cdots
\end{array}$$

'commutes', i.e. for each square the two routes between the top left and bottom right are in fact the same homomorphism.

We can carry over the terminology of simplicial complexes to the general situation and speak of *p-cycles* (elements of $Z_p = \text{Ker } d_p$), *p-boundaries* (elements of $B_p = \text{Im } d_{p+1}$) and *homology groups* $H_p = Z_p/B_p$, which exist because $d_{p-1} \circ d_p = 0$.

Thus a chain map takes cycles to cycles and boundaries to boundaries. It follows easily that there is an induced homomorphism

4.37
$$f_* : H_p \to H'_p$$
defined by $f_*(z + B_p) = f(z) + B'_p$ $(z \in Z_p)$

or, as we may write, using { } ambiguously,

$$f_*\{z\} = \{f(z)\}.$$

Examples 4.38. (1X) Taking, in the usual notation

$$C_p = C_p(L), \quad d_p = \partial_p : C_p(L) \to C_{p-1}(L)$$
$$C'_p = C_p(K), \quad d_p = \partial_p : C_p(K) \to C_{p-1}(K)$$

and $i_p : C_p(L) \to C_p(K)$ to be the inclusion, (i_p) is in fact a chain map, giving an induced homomorphism

$$i_* : H_p(L) \to H_p(K)$$

as in 4.30.

(2X) Taking

$$C_p = C_p(K), \quad d_p = \partial_p : C_p(K) \to C_{p-1}(K)$$
$$C'_p = C_p(K,L), \quad d_p = \widehat{\partial}_p : C_p(K,L) \to C_{p-1}(K,L)$$

and $j_p : C_p(K) \to C_p(K,L)$ to be the usual j_p (4.20), (j_p) is in fact a chain map giving an induced homomorphism

$$j_* : H_p(K) \to H_p(K,L)$$

as in 4.32.

(3X) Show that the 'restriction' ρ of the proof of 4.24(2) does not, except in special cases like $L = K$ or $L = \emptyset$, give a chain map between the appropriate chain complexes.

(4XX) Suppose that each f_p in 4.36 is an epimorphism (surjective). Show that

$$z' + f_p Z_p \to \{z'\} + \{f_* H_p\} \quad (z' \in Z'_p)$$

defines an isomorphism

$$Z'_p / f_p Z_p \to H'_p / f_* H_p.$$

(In particular this is true when $f_p = j_p$.)

(5X) Let $f_p : C_p(K,L) \to C_{p-1}(L)$ be the composite

$$C_p(K,L) \xrightarrow{\text{inclusion}} C_p(K) \xrightarrow{\partial_p} C_{p-1}(K) \xrightarrow{\rho} C_{p-1}(L).$$

Taking $C_p = C_p(K,L)$, $d_p = \widehat{\partial}_p$; $C'_p = C_{p-1}(L)$, $d'_p = \partial_{p-1}$ it is not quite true that (f_p) is a chain map from (C_p, d_p) to (C'_p, d'_q), for in fact $d'_p \circ f_p = -f_{p-1} \circ d_p$. Nevertheless f_p does take cycles to cycles and boundaries to boundaries, and there is an induced homomorphism $H_p(K,L) \to H_{p-1}(L)$, which is in fact the ∂_* of 4.34.

The notation for chain complexes and chain maps is easier to handle when the chain groups C_p are collected together into one big group C, as follows.

Definitions 4.39. Let (C_p, d_p) be a chain complex. The *direct sum* $C = \oplus\, C_p$ of the groups C_p is the abelian group whose elements are families $(c_p), c_p \in C_p, p \in \mathbb{Z}$, with all but a finite number of c_p equal to 0, and the operation $(c_p^1) + (c_p^2) = (c_p^1 + c_p^2)$. Thus if $C_p = 0$ when $p < 0$ or $p > n$, as is the case in the geometrical examples, then $C \cong C_0 \oplus C_1 \oplus \cdots \oplus C_n$ and an element of C can be regarded as an n-tuple $(c_0, c_1, \ldots, c_n)$.

The collection of homomorphisms (d_p) can be brought together into a single homomorphism $d : C \to C$, defined by: the element in the pth place of the family $d(c_p)$ is $d_{p+1} c_{p+1} \in C_p$.

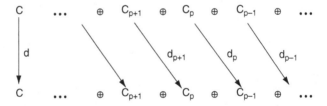

Then the condition $d_{p-1} \circ d_p = 0$ for all p is equivalent to $d \circ d = 0$, i.e. Im $d \subset$ Ker d. Also a chain map (f_p) can be thought of as a single homomorphism $f : C \to C'$ such that $d' \circ f = f \circ d$.

The group C, together with $d : C \to C$ is sometimes called a *differential graded group*, d being the differential and the decomposition of C as a direct sum being the grading. For further information see for example the books of Hilton & Wylie or Prasolov listed in the References. ★

5
The question of invariance

This chapter has a carefully chosen title: we are concerned here primarily with the question, and not with the answer. The question is this: suppose that K and K_1, are simplicial complexes triangulating the same object, i.e. with $|K| = |K_1|$. Is it true that $H_p(K) \cong H_p(K_1)$ for all p? The answer is in fact 'yes' – indeed a stronger assertion (5.13) is true, namely that the isomorphisms hold if $|K|$ is merely supposed homeomorphic to $|K_1|$, i.e. if there exists a continuous and bijective map $f:|K| \to |K_1|$. (The inverse map f^{-1} will automatically be continuous by a theorem of general topology, using the 'compactness' of $|K|$. See the books of Kelley or Lawson listed in the References.) The stronger assertion is referred to as the topological invariance of homology groups: it says that homology groups do not depend on particular triangulations but only on the underlying topological structure of $|K|$.

The topological invariance result can be restated in a negative form: if there exists an integer p such that $H_p(K)$ and $H_p(K_1)$ are not isomorphic, then $|K|$ and $|K_1|$ are not homeomorphic. Using results of Chapter 6 this in turn implies that no two inequivalent closed surfaces (2.8) have homeomorphic underlying spaces.

But we shall not prove the topological invariance theorem 5.13, since it would divert attention from the developments of succeeding chapters. The only invariance result actually needed is a much weaker one, and that one is proved in full; see 5.7.

Though this chapter up to 5.9 forms an integral part of the logical structure of the book, it could be omitted on a first reading. It might be a good idea to read the first part of the section 'Triangulations, simplicial approximation and topological invariance', say up to 5.10, which expands on the general statements above.

Invariance under stellar subdivision

Let K be a simplicial complex. A *subdivision* of K is a simplicial complex K_1 with the property that $|K_1| = |K|$ and given $s_1 \in K_1$ there exists $s \in K$ such that $s_1 \subset s$. Thus the simplexes of K_1 are contained in simplexes of K but K_1 and K triangulate the same subset of $\mathbb{R}^N$. An example is provided by the barycentric subdivision of a surface described following 2.6, where each triangle is divided into six by drawing the medians, and each edge is thereby halved; more of this anon. Another example is the stellar subdivision described in 3.29(5): a reader who does not want to work through that example should read what follows by specializing always to the case where K has dimension 2. That will suffice for the applications in this book and will give a perfectly adequate idea of the techniques involved. A *stellar subdivision* is the result of a sequence of subdivisions, each of these being *starring* of a simplex s at an interior point v. When K has dimension 2, the following pictures show the result of starring s at an interior point v. When s is a vertex no change takes place. In general, we remove $\overline{\mathrm{St}}(s, K)$ and replace it by the join $v \overset{\bullet}{s} \mathrm{Lk}(s, K)$; compare 3.24, 3.26.

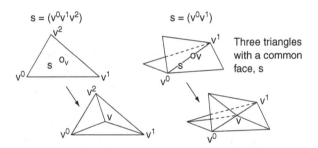

When K has dimension 2 we can produce the first barycentric subdivision of K by starring each 2-simplex at its centroid and then starring each 1-simplex of the original simplicial complex at its mid-point.

In fact the procedure of starring by descending dimension always produces the first barycentric subdivision; a general definition of barycentric subdivision is given in an appendix to this chapter for those who wish to verify this fact.

Invariance under stellar subdivision 129

Let K_1 be obtained from K by starring $s = (v^0 \ldots v^n)$ at v. Then all simplexes of K with s as a face are deleted, and certain simplexes, listed below, are introduced. If K is oriented, then K_1 can be oriented in a natural way, and the orientations are indicated in the list, where it is assumed that $(v^0 \ldots v^n)$ gives the correct orientation for s.

5.1 $\quad\begin{array}{l} (v), \text{ often written just } v \\ (v^0 \ldots v \ldots v^n) \text{ where } v \text{ replaces } v^i \text{ in } s \ (i = 0, \ldots, n) \\ \pm(vw^0 \ldots w^m) \\ \pm(v^0 \ldots v \ldots v^n w^0 \ldots w^m) \\ v \text{ replacing } v^i \text{ in } s \ (i = 0, \ldots, n) \end{array}\left.\begin{array}{l}\\ \\ \text{where } \pm (v^0 \ldots v^n w^0 \ldots w^m) \\ \text{(same sign as on the} \\ \text{left) belongs to } K \end{array}\right.$

The reader should check in some simple cases that these give the 'right' orientations for simplexes of K_1. Note that

$$(v^0 \ldots v \ldots v^n) = (-1)^i (v v^0 \ldots \widehat{v^i} \ldots v^n)$$

where v replaces v^i in s on the left of the equation.

Some notation will be useful in what follows. Write σ for the oriented simplex $(v^0 \ldots v^n)$, σ^i for $(v^0 \ldots \widehat{v^i} \ldots v^n)$, τ for $(w^0 \ldots w^m)$ and τ^j for $(w^0 \ldots \widehat{w^j} \ldots w^m)$. Joins are denoted as usual by concatenation of symbols ($\sigma\tau$ etc.) and $v\sigma$, for example, denotes $(v)\sigma = (vv^0 \ldots v^n)$. The notation can be extended to chains: for example $(\partial\sigma)\tau$ means $\Sigma(-1)^i(\sigma^i \tau) \in C_{m+n}(K)$.

Note that, in this notation, we have the following formula for the boundary of a join:

5.2 $\qquad \partial(\sigma\tau) = (\partial\sigma)\tau - (-1)^n \sigma(\partial\tau), \quad n = \dim \sigma.$

(If $\dim \sigma = 0$ then $(\partial\sigma)\tau$ is to be read as τ, and similarly if $\dim \tau = 0$ then $\sigma(\partial\tau) = \sigma$.)

Now we can define a homomorphism

$$\alpha_p \colon C_p(K) \to C_p(K_1)$$

as follows. Let $\sigma\tau = (v^0 \ldots v^n w^0 \ldots w^m)$ be a simplex in $\text{St}(s, K)$; then

$$\alpha_p(\sigma\tau) = v(\partial\sigma)\tau = \sum_{i=0}^{n}(v^0 \ldots v \ldots v^n w^0 \ldots w^m)$$

where v replaces v^i. This replaces $\sigma\tau$ by the sum of simplexes into which it is broken, with appropriate orientations. Also define $\alpha_p\sigma = v(\partial\sigma) = \sum(v^0 \ldots v \ldots v^n)$ (see diagram).

Other simplexes, not in St(s, K), α_p leaves unchanged.

Proposition 5.3. α_p *takes cycles to cycles and boundaries to boundaries and so induces a homomorphism*

$$\alpha_* : H_p(K) \to H_p(K_1)$$

defined by $\alpha_*\{z\} = \{\alpha_p z\}$.

Proof In fact the collection of homomorphisms α_p is a chain map (4.36), i.e. writing ∂ for the boundary homomorphism in either K or K_1, $\partial \circ \alpha_p = \alpha_{p-1} \circ \partial$ for all p. The result follows easily from this (compare 4.37). To prove the assertion about (α_p) it is enough to check that $\partial \circ \alpha_p(\sigma\tau) = \alpha_{p-1} \circ \partial(\sigma\tau)$ whenever $\sigma\tau \in K$, and also $\partial \circ \alpha_p \sigma = \alpha_{p-1} \circ \partial\sigma$. Using the formula for α_p above, 5.2 and the fact that $\partial \circ \partial = 0$ it follows without much difficulty that

$$\partial \circ \alpha_p(\sigma\tau) = \alpha_{p-1} \circ \partial(\sigma\tau) = (\partial\sigma)\tau - (-1)^n v(\partial\sigma)(\partial\tau)$$

where $n = \dim \sigma$. The other assertion follows similarly. $\square$

Next we shall define a homomorphism

$$\beta_p : C_p(K_1) \to C_p(K).$$

To do this we consider a more general situation.

Definition 5.4. Let K_1 and K be any two simplicial complexes. A *simplicial map*

$$\beta : K_1 \to K$$

is a map from the set of vertices of K_1 to the set of vertices of K with the property that

$$(w^0 \ldots w^p) \in K_1 \Rightarrow \{\beta w^0, \ldots, \beta w^p\} \text{ is the set of vertices}$$

of a simplex of K.

Note that $\beta w^0, \ldots, \beta w^p$ need not be distinct: for example β could map all vertices of K_1, to the same vertex of K.

Given a simplicial map $\beta : K_1 \to K$ as above we can define a homomorphism

$$\beta_p : C_p(K_1) \to C_p(K)$$

by

$$\beta_p(w^0 \ldots w^p) = \begin{cases} (\beta w^0 \ldots \beta w^p) & \text{if these } p+1 \text{ vertices} \\ & \text{of } K \text{ are distinct} \\ 0 & \text{otherwise} \end{cases}$$

Proposition 5.5. *With the above notation, β_p induces a homomorphism*

$$\beta_* : H_p(K_1) \to H_p(K)$$

defined by $\beta_*\{z\} = \{\beta_p z\}$.

Proof Again (β_p) is a chain map. The proof will be left to the reader, with the following hint: consider in turn the cases where $\beta w^0, \ldots, \beta w^p$ contains $p + 1$ distinct vertices of K, has one repetition, and has more than one repetition. In the first case $\beta_{p-1} \circ \partial(w^0 \ldots w^p) = \partial \circ \beta_p(w^0 \ldots w^p)$ since the formulae for the two sides are identical. In the other cases both sides are zero. $\square$

Returning to the case where K_1 is obtained from K by starring $s = (v^0 \ldots v^n)$ at v, there is a simplicial map $\beta : K_1 \to K$ defined by $\beta v = v^0$ and $\beta w = w$ for $w \neq v$. By 5.5 this gives a homomorphism $\beta_* : H_p(K_1) \to H_p(K)$.

Theorem 5.6. *With the notation just given, and that of 5.3, α_* and β_* are inverse isomorphisms, so that*

$$\alpha_* : H_p(K) \cong H_p(K_1).$$

Proof It is immediate from the definitions that $\beta_p \circ \alpha_p = $ identity; hence $\beta_* \circ \alpha_*\{z\} = \beta_*\{\alpha_p z\} = \{\beta_p \circ \alpha_p z\} = \{z\}$ for any $\{z\} \in H_p(K)$, i.e. $\beta_* \circ \alpha_* = $ identity.

Next note that, for any $c \in C_p(K_1)$, $\alpha_p \circ \beta_p c - c$ contains only simplexes in $\overline{\text{St}}(v, K_1)$, since β_p does not change anything outside $\overline{\text{St}}(v, K_1)$, and α_p does not change anything outside $\overline{\text{St}}(s, K)$. If $z \in Z_p(K_1)$, it follows that $\alpha_p \circ \beta_p z - z$ is a cycle on $\overline{\text{St}}(v, K_1)$, since α_p and β_p take cycles to cycles. Now $\overline{\text{St}}(v, K_1)$ is the join of $\{v\}$ to the link $\text{Lk}(v, K_1)$ (see 3.29(4)): i.e. the closed star is a cone on the link. Now any p-cycle ($p > 0$) on a cone is homologous to zero. This follows at once from 6.8, or alternatively from the following direct argument.

Let $z' \in Z_p(\overline{\mathrm{St}}(v, K_1))$ and write L for $\mathrm{Lk}(v, K_1)$. Then there are chains $c' \in C_{p-1}(L)$ and $c'' \in C_p L$ such that $z' = vc' + c''$ (thus c' contains the simplexes in z' which have v as a vertex). Hence $0 = \partial z' = c' - v(\partial c') + \partial c''$ from 5.2, and this implies that $c' + \partial c'' = 0$ and $\partial c' = 0$. Hence

$$\partial(vc'') = c'' - v(\partial c'') = c'' + vc' = z'$$

so that $z' \sim 0$.

This shows that $\alpha_p \circ \beta_p z \sim z$, and so $\alpha_* \circ \beta_* =$ identity for $p > 0$. For $p = 0$, we have $\alpha_0 \circ \beta_0 v - v = v^0 - v \sim 0$ (since $(v^0 v) \in K_1$) and $\alpha_0 \circ \beta_0 w = w$ for $w \neq v$; hence the result, namely $\alpha_* \circ \beta_* =$ identity for all p. □

Corollary 5.7 (Invariance under stellar subdivision). *Suppose that K_1 is a stellar subdivision of K, i.e. is obtained from K by a sequence of starrings. Then $H_p(K_1) \cong H_p(K)$ for all p. In particular the isomorphism holds if K_1 is obtained from K by one or more barycentric subdivisions.*

Proof This follows immediately by repeated application of the theorem. □

5.8. The relative case In order to avoid overburdening the above proofs with notation, the absolute case only was considered. Nevertheless the same arguments can be used for a pair (K, L) consisting of a simplicial complex K and a subcomplex L. Thus (K_1, L_1) is a subdivision of (K, L) if K_1 subdivides K and L_1 subdivides L. A stellar subdivision K_1 of K will automatically give a stellar subdivision L_1 of L, and in the case of a single starring the homomorphisms α_* and β_* can be defined as before, but this time between $H_p(K, L)$ and $H_p(K_1, L_1)$. (Note that a simplicial map $\beta : (K_1, L_1) \to (K, L)$ is a simplicial map $K_1 \to K$ such that $(w^0 \ldots w^p) \in L_1 \Rightarrow \{\beta w^0, \ldots, \beta w^p\}$ are the vertices of a simplex of L.) As before, α_* and β_* are inverse isomorphisms, and it follows that $H_p(K_1, L_1) \cong H_p(K, L)$ whenever (K_1, L_1) is obtained from (K, L) by successive starrings.

Remark 5.9. Two simplicial complexes K and L are called *piecewise linearly (p.l.) equivalent* if there are subdivisions K_1 of K and L_1 of L such that K_1 and L_1 are isomorphic. For example, any two closed surfaces with the same standard form are p.l. equivalent (as well as being equivalent in the sense of 2.8). To see this informally superimpose their standard polygons (assumed regular) and subdivide to make a common subdivision of each; the diagram indicates roughly how this is done for two overlapping triangles. The converse is also true: if two surfaces are p.l. equivalent then they have the same standard form. Compare 2.16(8), which says that Euler characteristic is unaltered by subdivision.

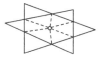

The same sort of subdivision construction as that illustrated above can be used to show that any two triangulations of the same polyhedron, i.e. any two simplicial complexes in $\mathbb{R}^N$ with the same underlying space, have a common subdivision, and hence are p.l. equivalent.

★ Triangulations, simplicial approximation and topological invariance

In this section we shall sketch very briefly the route to a more general invariance theorem than 5.7. Full details appear in many textbooks of algebraic topology, for example the books by Prasolov *Elements of Homology Theory* or Hilton & Wylie listed in the References. There is also a pleasant account of the two-dimensional case in the book of Ahlfors & Sario.

Let X be a subset of a euclidean space $\mathbb{R}^N$. Then certainly X may fail to be the underlying space of any simplicial complex – for example if X is a circle – but the next best thing to hope for is that there is some simplicial complex K such that $|K|$ is homeomorphic to X. Thus taking X to be the unit circle in $\mathbb{R}^2$ and K to be the set of edges and vertices of the square in $\mathbb{R}^2$ with vertices $\left(\pm\frac{\sqrt{2}}{2}, \pm\frac{\sqrt{2}}{2}\right)$ a homeomorphism $f : |K| \to X$ can be defined by radial projection from the origin.

Let $X \subset \mathbb{R}^N$. Given a simplicial complex K and a homeomorphism $h : |K| \to X$ we call K (or, more precisely, h) a *triangulation* of X. In this sense the closed surfaces of Chapter 2 are triangulations of appropriate 'curved surfaces' in euclidean space. For example the hollow tetrahedron is a triangulation of a 2-sphere, h being given say by radial projection.

134   The question of invariance

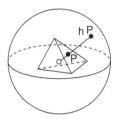

As another example, the unit disk D^2 in $\mathbb{R}^2$ is the set of points (x, y) with $x^2 + y^2 \leqslant 1$; thus a triangulation of D^2 is a homeomorphism $h : |K| \to D^2$ where K is a simplicial complex. Now K must be two-dimensional (since h is a homeomorphism); the image of the 1-skeleton of K in D^2 will be a curved graph in the sense of the discussion leading up to Fáry's theorem 1.34. By that theorem there is a homeomorphism $f : \mathbb{R}^2 \to \mathbb{R}^2$ taking this curved graph to a straight graph, so that $f \circ h$ defines an isomorphism between K and a simplicial complex in $\mathbb{R}^2$. Thus every triangulation of the disk D^2 can be realized in $\mathbb{R}^2$. With slightly more trouble one can prove, also using Fáry's theorem, that every triangulation of the 2-sphere $\{(x, y, z) \in \mathbb{R}^3 : x^2 + y^2 + z^2 = 1\}$ can be realized in $\mathbb{R}^3$.

When K is a triangulation of X we cannot immediately define the homology groups of X to be those of K, since there could well be another simplicial complex K_1 and a homeomorphism $h_1 : |K_1| \to X$; we do not know that $H_p(K_1) \cong H_p(K)$ for all p. As an example, K_1 could be any subdivision of K, in which case h_1 could equal h, and then we know the existence of an isomorphism only for stellar subdivisions.

What is needed is a result which asserts the following: Suppose that K_1 and K are oriented simplicial complexes and that $f : |K_1| \to |K|$ is a homeomorphism. Then $H_p(K_1) \cong H_p(K)$ for all p. Granted this and given h_1 and h as above, we simply take $f = h^{-1} \circ h_1$ to show that the two triangulations of X have isomorphic homology groups. There follows a brief summary of the steps needed to obtain the above result.

It is worth noting that it is not necessarily true that K_1 and K as above are p.l. equivalent (5.9). At one time it was conjectured that any two triangulations of homeomorphic polyhedra were p.l. equivalent; this was known as the Hauptvermutung (main conjecture). The first counterexample (six-dimensional polyhedra) was found by Milnor (1961), and subsequently it was discovered that there exist 'topological manifolds' of any dimension ≥ 4 which do not admit of any triangulation, and other four-dimensional manifolds which admit of infinitely many distinct p.l. structures. This area is still the subject of research in topology. See Freedman (1982), and also the book edited by Ranicki listed in the References.

Let K_1 and K be any two simplicial complexes and let $f : |K_1| \to |K|$ be a continuous map. We 'approximate' f by a simplicial map $\beta : K_1 \to K$.

Definition 5.10. A simplicial map $\beta : K_1 \to K$ is called a *simplicial approximation* to $f : |K_1| \to |K|$ provided that, for each vertex v of K_1,

$$f(\mathrm{St}(v, K_1)) \subset \mathrm{St}(\beta v, K).$$

As an example, the diagram shows a hexagon K_1 (vertices $1, \ldots, 6$) and a triangle K (vertices 1,3,5) inscribed in a circle. A continuous map $|K_1| \to |K|$ is defined by projection towards the centre of the circle, and a simplicial approximation to this is $\beta : K_1 \to K$ where β takes 1,2,3,4,5,6 to, respectively, 1,3,3,5,5,1. On the other hand consider the inverse map $g : |K| \to |K_1|$ given by outward projection. There is no simplicial map $K \to K_1$ which is a simplicial approximation to g, since $g(\mathrm{St}(1, K))$ is not contained in the star of any vertex of K_1. However K can be barycentrically subdivided – once is enough, giving K' – so that the map $g : |K'| = |K| \to |K_1|$ does have a simplicial approximation. In general, the following result holds.

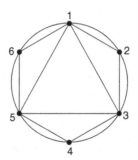

5.11. Simplicial approximation theorem *Given simplicial complexes K_1 and K and a continuous map $f : |K_1| \to |K|$ there exists an integer r and a simplicial map $\beta : K_1^{(r)} \to K$ (the superscript denoting rth barycentric subdivision) which is a simplicial approximation to f.* □

Suppose, then, that K_1, K and f are as in 5.11; by 5.7 there is an isomorphism $\gamma_* : H_p(K_1) \to H_p(K_1^{(r)})$ and so a homomorphism

$$f_* : H_p(K_1) \xrightarrow{\gamma_*} H_p(K_1^{(r)}) \xrightarrow{\beta_*} H_p(K)$$

for each p. One vital, and far from obvious, fact about f_* is that it does not depend on r or β, so long as β is a simplicial approximation to f. Thus it really depends only on f, and the notation f_* is justified. Another important fact is this:

Lemma 5.12. *Let* $g : |K_2| \to |K_1|$ *and* $f : |K_1| \to |K|$ *be continuous maps, where* K_2, K_1 *and* K *are (oriented) simplicial complexes. Then* $(f \circ g)_* = f_* \circ g_* :$ $H_p(K_2) \to H_p(K)$ *for each* p. □

The result required now follows at once by taking $K_2 = K$ and f a homeomorphism with inverse g. Hence:

5.13. Topological invariance of homology groups *Let* K_1 *and* K *be (oriented) simplicial complexes and* $f : |K_1| \to |K|$ *be a homeomorphism. Then* $f_* :$ $H_p(K_1) \to H_p(K)$ *is an isomorphism for each* p. □

★

★ Appendix on barycentric subdivision

Let s be an n-simplex, $s = (v^0 \ldots v^n)$. The *barycentre* of s is the point $\widehat{s} = \frac{1}{n+1}(v^0 + v^1 + \cdots + v^n)$. Thus the barycentre of a 0-simplex (v) is v, the barycentre of a 1-simplex $(v^0 v^1)$ is its mid-point, and the barycentre of a 2-simplex $(v^0 v^1 v^2)$ is its centroid.

The *first barycentric subdivision* K' of a simplicial complex K is the simplicial complex with vertex set $\{\widehat{s} : s \in K\}$, and with simplexes $t = (\widehat{s^0}\widehat{s^1} \ldots \widehat{s^q})$ where $s^0 < s^1 < \cdots < s^q$ and $s^0, \ldots, s^q$ are simplexes of K. (The superscript is just for indexing, and does not necessarily give the dimension of s^i.) Notice that $t \subset s^q$; to see that K' is a subdivision of K we must therefore show that $|K| \subset |K'|$. If $x \in |K|$ then $x \in s$ for some $s \in K$ – we can take the s of lowest dimension for which this is true, so that if $s = (v^0 \ldots v^p)$ then $x = \lambda_0 v^0 + \cdots + \lambda_p v^p$ where the barycentric coordinates are all > 0. We can also assume $\lambda_0 \geqslant \lambda_1 \geqslant \cdots \geqslant \lambda_p$. Let $s^i = (v^0 \ldots v^i)$ $(i = 0, \ldots, p)$; then it is straightforward to check that $x \in (\widehat{s^0}\widehat{s^1} \ldots \widehat{s^p})$: in fact

$$x = (\lambda_0 - \lambda_1)\widehat{s}^0 + 2(\lambda_1 - \lambda_2)\widehat{s}^1$$
$$+ \cdots + p(\lambda_{p-1} - \lambda_p)\widehat{s}_{p-1} + (p+1)\lambda_p\widehat{s}_p.$$

If L is a subcomplex of K then L is automatically subdivided into L' when K is subdivided into K'. The reader may like to check the following assertions.

(1) $\widehat{s}$ is a vertex of L' if and only if $s \in L$.
(2) If t is a simplex of K' such that all vertices of t belong to L', then $t \in L'$.

(3) Suppose that $(vw) \in K'$, where v is a vertex of L' and w is not a vertex of L'; then w is not a vertex of K (hence is the barycentre of a simplex of K of dimension > 0).

(4) Suppose that $s \in K$ and that no vertex of s belongs to L. Then there is no edge $(vw) \in K'$ with $v \in L'$ and w the barycentre of a face of s. ★

6
Some general theorems

In this chapter and the next we shall prove some general results which will find application in Chapter 9. For the present the motivation lies in the evident difficulty of calculating homology groups directly from the definition, and the consequent need for some help from general theorems. Enough is proved here to enable us to calculate the homology groups of all the closed surfaces described in Chapter 2, without difficulty. We shall also calculate the homology groups of some other standard simplicial complexes such as cones (6.8) and spheres (6.11(3)), and show (6.10) that relative homology groups can be defined in terms of ordinary homology groups.

The method presented here for the calculation of homology groups of closed surfaces is based on the idea of collapsing which was introduced in 3.30. This is not the only available method. From the point of view of this book, the advantage of using it is that, having established the invariance of homology groups under barycentric subdivision (5.7), we are in possession of a rapid and rigorous approach to the results. Another method of wide application comes from the Mayer–Vietoris sequence, presented in Chapter 7, where indications are given as to how the homology groups of closed surfaces can be re-calculated by use of the sequence. For effective and unfettered use of the Mayer–Vietoris sequence one really needs to use something like the topological invariance theorem stated in 5.13 – that is, it is better to forget about particular triangulations altogether. Some uninhibited calculations using the Mayer–Vietoris sequence appear in the next chapter, starting at 7.5.

The homology sequence of a pair

Let K be an oriented simplicial complex of dimension n and L a subcomplex of K. Recall from 4.30, 4.32 and 4.34 the homomorphisms i_*, j_* and ∂_*; the

The homology sequence of a pair

following sequence is called the *homology sequence of the pair* (K,L).

$$0 \to H_n(L) \xrightarrow{i_*} H_n(K) \xrightarrow{j_*} H_n(K,L) \xrightarrow{\partial_*} H_{n-1}(L) \xrightarrow{i_*} H_{n-1}(K) \xrightarrow{j_*} \cdots$$

$$\cdots \xrightarrow{j_*} H_1(K,L) \xrightarrow{\partial_*} H_0(L) \xrightarrow{i_*} H_0(K) \xrightarrow{j_*} H_0(K,L) \to 0$$

Theorem 6.1. *The homology sequence of the pair (K,L) is exact.*

Proof As usual, we regard $C_p(L)$ as a subgroup of $C_p(K)$ (see the note preceding 4.24), and denote by $\rho : C_p(K) \to C_p(L)$ the 'restriction' which puts coefficients of simplexes not in L equal to 0. We have to prove exactness at three places – $H_p(L), H_p(K), H_p(K,L)$ – and each of these involves two proofs of inclusion.

(1) *Exactness at $H_p(L)$.* Let $z \in Z_{p+1}(K,L)$; then $i_* \partial_* \{z\}_{K,L} = \{\partial z\}_K$. However $z \in C_{p+1}(K)$, so $\partial z \in B_p(K)$, i.e. $\{\partial z\}_K = 0$. Hence $\operatorname{Im} \partial_* \subset \operatorname{Ker} i_*$.
Suppose, conversely, that $w \in Z_p(L)$ and $i_*\{w\}_L = \{w\}_K = 0$. Then $w = \partial u$ for some $u \in C_{p+1}(K)$. Writing $u = ju + \rho u$ we have

$$w = \partial u = \partial j u + \partial \rho u.$$

Hence $\partial ju \in C_p(L)$, that is, $ju \in Z_{p+1}(K,L)$. Since $\partial_*\{ju\}_{K,L} = \{\partial ju\}_L = \{w\}_L - \{\partial \rho u\}_L = \{w\}_L$ ($\partial \rho u \in B_p(L)$), we have $\{w\}_L \in \operatorname{Im} \partial_*$, so $\operatorname{Ker} i_* \subset \operatorname{Im} \partial_*$. Notice that although ∂ju may not equal w (see the left-hand figure), it is homologous to w, since $\partial \rho u \sim 0$ in L.

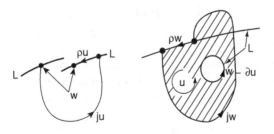

(2) *Exactness at $H_p(K)$.* Let $z \in Z_p(L)$; then $j_* i_*\{z\}_L = \{jz\}_{K,L} = \{0\} = 0$. Hence $\operatorname{Im} i_* \subset \operatorname{Ker} j_*$. (This is the only one of the three cases where the composite homomorphism is trivial even at the chain level.)
Conversely let $w \in Z_p(K)$ be such that $j_*\{w\}_K = \{jw\}_{K,L} = 0$. Then (see the right-hand figure) $jw = \widehat{\partial u} = j\partial u$ for some $u \in C_{p+1}(K,L)$. Hence $w - \partial u \in C_p(L)$; indeed $\partial(w - \partial u) = \partial w = 0$ so $w - \partial u \in Z_p(L)$.

Finally $i_*\{w - \partial u\}_L = \{w - \partial u\}_K = \{w\}_K$ (since $\partial u \in B_p(K)$), so that $\{w\}_K \in \text{Im } i_*$. This shows $\text{Ker } j_* \subset \text{Im } i_*$.

(3) *Exactness at* $H_p(K, L)$. Let $z \in Z_p(K)$; then $\partial_* j_*\{z\}_K = \{\partial j z\}_L = \{\partial(z - \rho z)\}_L = \{-\partial \rho z\}_L$ ($\partial z = 0$) = 0 since $\rho z \in C_p(L)$. (Same diagram as in (2), with z replacing w.) Hence $\text{Im } j_* \subset \text{Ker } \partial_*$.

Conversely, let $w \in Z_p(K, L)$ be such that $\partial_*\{w\}_{K,L} = \{\partial w\}_L = 0$. Thus $\partial w = \partial u$, for some $u \in C_p(L)$, so that $\partial(w - u) = 0 : w - u \in Z_p(K)$. Further $j_*\{w - u\}_K = \{jw - ju\}_{K,L} = \{w\}_{K,L}$. Thus $\{w\}_{K,L} \in \text{Im } j_*$; hence $\text{Ker } \partial_* \subset \text{Im } j_*$.

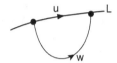

□

Remark 6.2. There is a slightly different form of the sequence in 6.1 which is sometimes more convenient on account of the fact that it contains reduced zero-dimensional homology groups, and these are zero for connected simplicial complexes (1.23, 4.9(2)). Recall (4.8), $\widetilde{H}_0(K) = \text{Ker } \varepsilon_K / \text{Im } \partial_1$, where $\varepsilon_K : C_0(K) \to \mathbb{Z}$ is the augmentation. For $z \in \text{Ker } \varepsilon_K$ write $[z] = [z]_K = z + \text{Im } \partial_1 \in \widetilde{H}_0(K)$.

$$\begin{aligned}
\text{Define } \widetilde{\partial}_* &: H_1(K, L) \to \widetilde{H}_0(L) \\
\text{by} & \quad \{z\}_{K,L} \to [\partial z]_L \\
\widetilde{i}_* &: \widetilde{H}_0(L) \to \widetilde{H}_0(K) \\
\text{by} & \quad [z]_L \to [z]_K \\
\text{and } \widetilde{j}_* &: \widetilde{H}_0(K) \to H_0(K, L) \\
\text{by} & \quad [z]_K \to \{jz\}_{K,L}.
\end{aligned}$$

It is left to the reader to check that these really do define homomorphisms, and that the *reduced sequence of the pair* (K, L) identical with the above sequence but ending

$$\cdots \to H_1(K, L) \xrightarrow{\widetilde{\partial}_*} \widetilde{H}_0(L) \xrightarrow{\widetilde{i}_*} \widetilde{H}_0(K) \xrightarrow{\widetilde{j}_*} \widetilde{H}_0(K, L) \to 0$$

is exact provided L is non-empty.

Examples 6.3. (1) The exact sequence of (K, L) can sometimes be used to derive the relative homology groups of (K, L) from knowledge of the homology groups of L, those of K, and the homomorphisms i_*. The technique is to get

$H_p(K, L)$ in the middle of a short exact sequence:

$$0 \to \operatorname{Ker} \partial_* \to H_p(K, L) \xrightarrow{\partial_*} \operatorname{Im} \partial_* \to 0$$

$$\| \qquad\qquad\qquad\qquad\qquad \|$$

$$\operatorname{Im} j_* \qquad\qquad\qquad \operatorname{Ker}\left(i_* : H_{p-1}(L) \to H_{p-1}(K)\right)$$

$$?\|$$

$$H_p(K)/\operatorname{Im}(i_* : H_p(L) \to H_p(K))$$

If this sequences splits (A.19) then $H_p(K, L)$, including generators, can be calculated – see the examples below.

(2) Let K and L be as in 4.29(3) (Möbius band and its rim). Using the calculations of 4.15, the groups and homomorphisms written in under the following reduced homology sequence are known:

$$H_2(L) \to H_2(K) \to H_2(K,L) \to H_1(L) \xrightarrow{i_*} H_1(K) \xrightarrow{j_*} H_1(K,L)$$
$$\quad 0 \qquad\quad 0 \qquad\qquad\qquad\qquad \mathbb{Z} \xrightarrow{\times 2} \mathbb{Z}$$

$$\to \widetilde{H}_0(L) \to \widetilde{H}_0(K) \to H_0(K,L) \to 0$$
$$\quad\; 0 \qquad\quad\; 0$$

Hence
$H_2(K, L) \cong \operatorname{Ker} i_* = 0$,
$H_1(K, L) \cong H_1(K)/\operatorname{Im} i_* \cong \mathbb{Z}_2$, with generator the image under j_* of a generator of $H_1(K)$,
$H_0(K, L) = 0$.
(If we use the unreduced sequence it is only necessary to see that i_* : $H_0(L) \to H_0(K)$ is an isomorphism (compare 4.31(3)), which implies that the homomorphisms before and after this i_* are trivial.)

(3X) Use the (unreduced) homology sequence to calculate the relative homology groups of (cylinder, both ends) – see 4.29(1) for diagram. In this case i_*: $H_0(L) \to H_0(K)$ has infinite cyclic kernel, generated by $\{v^1 - v^2\}_L$. Also $i_* : H_1(L) \to H_1(K)$ takes each generator $\{e^7 + e^8 + e^9\}_L, \{e^{10} + e^{11} + e^{12}\}_L$ to a generator of $H_1(K)$; in particular this i_* is epic (surjective). Use the method of (1) above.

(4X) Take $K =$ the cylinder of 4.29(1) and $L =$ one end. This time the reduced homology sequence of (K, L) (as in (2) above) and the fact that $i_* : H_1(L) \to H_1(K)$ is an isomorphism show $H_2(K, L)$ and $H_1(K, L)$ are both trivial.

(5X) Show that, for any (K, L),

$H_p(K, L) = 0$ for all $p \Leftrightarrow i_* : H_p(L) \to H_p(K)$ is an isomorphism for all p.

(6) Let K^1 be the 1-skeleton of K; then $H_1(K^1) = Z_1(K^1) = Z_1(K)$ by 4.9(1) and (3). Similar remarks apply to L and (K,L); applying 6.1 to the pair (K^1, L^1) we obtain an exact sequence

$$0 \to Z_1(L) \to Z_1(K) \to Z_1(K,L) \to H_0(L) \to H_0(K) \to H_0(K,L) \to 0.$$

This was used in the proof of 4.27.

(7X) Let K be a connected oriented graph and L a sub-complex of K with components $L_1, \ldots, L_\ell$. Write $\mu(L_\alpha) = \mu_\alpha$ ($\alpha = 1, \ldots, \ell$), thus $\mu_\alpha = $ rank $Z_1(L_\alpha) = $ rank $H_1(L_\alpha)$; and $\mu(K) = \mu$. Noting that $H_1(K,L)$ is free abelian (4.23(1)), show that $H_1(K,L)$ has rank $\geq \ell - 1$, and that, if the rank is $\ell - 1$ then $\mu = \mu_1 + \cdots + \mu_\ell$. (Compare 4.27.) Also, if $\mu = \mu_1 + \cdots + \mu_n$, then $i_* : H_1(L) \to H_1(K)$ is an isomorphism.

(8X) In the example of 4.31(4), calculate $H_1(K,L)$ and $H_2(K,L)$ from the reduced homology sequence of (K,L). (First verify that $H_2(K) = 0$.)

(9X) Let T be a maximal tree in the 1-skeleton of a connected simplicial complex K. Use the reduced homology sequence of (K,T) to prove that

$$j_* : H_1(K) \to H_1(K,T)$$

is an isomorphism. Show that the edges of K not in T form a basis for $Z_1(K,T)$. Thus if $e^1, \ldots, e^\mu$ are these edges, a presentation of $H_1(K,T)$ is

$$\left(\begin{matrix} e^1, \ldots, e^\mu \\ \widehat{\partial} t = 0 \quad \text{for each 2-simplex } t \text{ of } K. \end{matrix} \right)$$

Given generators for $H_1(K,T)$ how does one obtain generators for $H_1(K)$? What is the connexion between this method and that of 4.14?

(10X) Use 4.38(4) to show that, in the usual notation,

$$j(Z_p(K)) = Z_p(K,L) \Leftrightarrow i_* : H_{p-1}(L) \to H_{p-1}(K) \text{ is injective.}$$

Taking $p = 1$, compare with 4.29(4) and 4.31(3).

The excision theorem

The next theorem will be used mainly in Chapter 9, though a few applications will be given here. It derives its name from the fact that it compares the relative homology groups of a pair $(L_1 \cup L_2, L_1)$ with those of a pair $(L_2, L_1 \cap L_2)$ obtained by 'excising' $L_1 \setminus L_2$.

The excision theorem

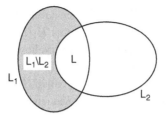

Let K be an oriented simplicial complex and L_1, L_2 subcomplexes of K with $K = L_1 \cup L_2$. Write $L = L_1 \cap L_2$. As usual we regard all chain groups as subgroups of those of K.

Theorem 6.4. $H_p(K, L_1) \cong H_p(L_2, L)$ *for all p (we could write $=$ here).*

Proof The groups $C_p(K, L_1)$ and $C_p(L_2, L)$ are clearly the same subgroup of $C_p(K)$. That is not quite the end of the proof, because we must check that the relative boundary homomorphisms

$$C_p(K, L_1) \to C_{p-1}(K, L_1) \quad \text{and} \quad C_p(L_2, L) \to C_{p-1}(L_2, L)$$

coincide. The result will then follow from the definition of relative homology groups. Let

$$j : C_{p-1}(K) \to C_{p-1}(K, L_1) \quad \text{and} \quad j' : C_{p-1}(L_2) \to C_{p-1}(L_2, L)$$

be the usual homomorphisms: in each case the homomorphism puts at zero the coefficient of any simplex not in L_1, so that if $c \in C_{p-1}(L_2)$ then $jc = j'c \in C_{p-1}(K, L_1) = C_{p-1}(L_2, L)$. Hence, for any $c' \in C_p(K, L_1) = C_p(L_2, L)$, we have $j\partial c' = j'\partial c'$. This *is* the end of the proof. □

Example 6.5. Let M be a two-dimensional simplicial complex and s a 2-simplex of M. Then removing s from M leaves a subcomplex K say. We ask the following question: when does $H_1(K)$ depend, up to isomorphism, only on M and not on the particular 2-simplex s which was removed? Now $M = K \cup \bar{s}$, $\overset{\bullet}{s} = K \cap \bar{s}$, where, as usual, $\bar{s}$ and $\overset{\bullet}{s}$ are the set of faces and proper faces of s, respectively. By excision $H_p(M, K) = H_p(\bar{s}, \overset{\bullet}{s})$, which is isomorphic to $\mathbb{Z}$ if $p = 2$ and to 0 if $p = 1$, by a straightforward calculation (or see 4.23(4)).

Consider the exact sequence (6.1)

$$H_2(K) \to H_2(M) \to H_2(M, K) \to H_1(K) \to H_1(M) \to H_1(M, K)$$
$$ \mathbb{Z} 0$$

144 Some general theorems

If $H_1(M)$ is free abelian and $H_2(M) = 0$ then it follows from the exact sequence that $H_1(K) \cong H_1(M) \oplus \mathbb{Z}$ and therefore that $H_1(K)$ does not depend on the choice of s. These conditions are not necessary, however, for if M is a closed surface the same results holds (this is proved in the course of proving 6.13). (XX) The reader is invited to find an M for which $H_1(K)$ does depend on s and to explain this dependence from the exact sequence.

Collapsing revisited

The idea of collapsing was introduced in 3.30, but the reason for introducing it is the following result.

Theorem 6.6. *Suppose that an oriented simplicial complex K collapses to a subcomplex L. Then the homomorphism*

$$i_* : H_p(L) \to H_p(K)$$

is an isomorphism for all p.

Proof Suppose first that K collapses to L by an elementary collapse, so that $K = L \cup \{s, t\}$ say where s is a principal q-simplex of K and t is a free face of s.

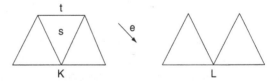

We shall prove that $H_p(K, L) = 0$ for all p; the result then follows immediately from the homology sequence of (K, L).

The chain complex of (K, L) contains only two non-zero relative chain groups, namely those in dimensions q and $q - 1$, and therefore takes the form

$$\cdots \to 0 \to C_q(K, L) \xrightarrow{\widehat{\partial}} C_{q-1}(K, L) \to 0 \to \cdots$$

Now $C_q(K, L)$ is infinite cyclic, generated by the oriented simplex σ corresponding to s, and $C_{q-1}(K, L)$ is also infinite cyclic, generated by the oriented simplex τ corresponding to t. Furthermore $\widehat{\partial}\sigma = \pm\tau$ so that $\widehat{\partial}$ is an isomorphism. It follows from the definition of relative homology groups that $H_p(K, L) = 0$ for all p.

For the general case, consider a sequence of elementary collapses $K \searrow^e K_1 \searrow^e K_2 \searrow^e \cdots \searrow^e K_m = L$. It is straightforward to verify that the homomorphism

$$i_* : H_p(L) \to H_p(K)$$

is the composite

$$H_p(K_m) \to H_p(K_{m-1}) \to \cdots \to H_p(K_1) \to H_p(K)$$

where each homomorphism is given by the inclusion of a subcomplex. Consequently i_* is an isomorphism by the first part. □

Note that this theorem does not just tell us $H_p(K)$ up to isomorphism: it says that p-cycles whose homology classes on L generate $H_p(L)$ have homology classes on K which generate $H_p(K)$.

Examples 6.7. (1) The results in 4.15 and 4.17(1) (the homology groups of Möbius band and cylinder) become obvious, using 6.6, since each simplicial complex collapses to a simple closed polygon.

(2) As a special case of 6.6, if K collapses to a point, then $H_p(K) = 0$ ($p > 0$); $H_0(K) \cong \mathbb{Z}$. If K is a graph, then the converse is true – see 1.12(2). However there exist two-dimensional simplicial complexes which satisfy the homological condition just given but which cannot be collapsed at all since there is no principal simplex with a free face. An example is given by any triangulation of the 'dunce hat'; see 3.22(4).

Before going on to calculate the homology groups of closed surfaces by collapsing, here is a simple application to cones.

Let K be a simplicial complex in $\mathbb{R}^N$ and let $v \in \mathbb{R}^N$ be a vector not affinely dependent on the vertices of K (if necessary add one to N in order to find such a v). The *cone on K with vertex v*, denoted CK, is the simplicial complex whose simplexes are of the form

$$(v) \quad \text{or} \quad (vv^0 \ldots v^p) \quad \text{or} \quad (v^0 \ldots v^p)$$

where $(v^0 \ldots v^p) \in K$. (K is called the *base* of the cone.) Thus CK is the join of K to the simplicial complex $\{v\}$ (see 3.26), and, up to isomorphism, is independent of the choice of v. If K is oriented then CK can be oriented by regarding the above elements of CK as oriented simplexes.

Theorem 6.8. (Homology groups of a cone) *With the above notation,*

$$H_p(CK) = 0 \quad \text{for} \quad p > 0$$

while $\quad H_0(CK) \cong \mathbb{Z}.$

Proof We prove CK collapses to a point – in fact to v – and the result then follows from 6.6. If K is empty then $CK = \{v\}$, so there is nothing to collapse (or to prove). Otherwise we proceed by induction on the dimension of K. If $\dim K = n$, then any $(n + 1)$-simplexes of CK can be collapsed, removing the n-simplexes of K and therefore leaving a cone on a simplicial complex of dimension $n - 1$. If $\dim K = 0$ the same procedure collapses CK directly to the vertex of the cone. □

Corollaries 6.9. (1) *Suppose that L is a subcomplex of K and that L is a cone. Then*

$$j_* : H_p(K) \to H_p(K, L) \quad p > 0$$

and $\quad \tilde{j}_* : \tilde{H}_0(K) \to H_0(K, L) \quad \text{(see 6.2)}$

are isomorphisms.

(2) *For any K,*

$$H_p(CK, K) \cong H_{p-1}(K) \quad p > 1$$
$$H_1(CK, K) \cong \tilde{H}_0(K)$$

(3) (*Homology groups of a simplex*) *Let s be an n-simplex $(n \geqslant 0)$ and $K = \bar{s}$ (oriented arbitrarily). Then $H_p(K) = 0$ for $p > 0$ and $H_0(K) \cong \mathbb{Z}$.*

Proof (1) The homology sequence of the pair (K, L) contains the section

$$H_p(L) \to H_p(K) \xrightarrow{j_*} H_p(K, L) \to H_{p-1}(L).$$

For $p > 1$, both end-groups are zero, by 6.8, and the result follows from exactness. For $p = 0, p = 1$ the result follows from a similar application of the reduced homology sequence.

(2) The proof is similar to that of (1).

(3) K is in fact a cone, with empty base if $n = 0$ and with base the closure of an $(n - 1)$-dimensional face of s if $n > 0$. Thus the result follows from 6.8. □

Using the excision theorem and 6.8 we can establish the following interpretation of relative homology groups as the absolute homology groups of an associated simplicial complex. The idea is that the relative homology groups of the pair (K, L) in some sense disregard L, and homologically this can be done by adding a cone on L. Thus let $K \cup CL$ be obtained by taking the union of K and a cone CL on L, the cone meeting K precisely on L as in the sketch.

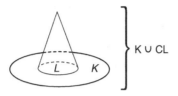

Theorem 6.10. *With the above notation*

$$H_p(K, L) \cong H_p(K \cup CL) \quad p > 0$$
$$\text{and} \quad H_0(K, L) \cong \tilde{H}_0(K \cup CL).$$

(N.B. If L is empty, then CL is a single point not in $|K|$.)

Proof By the excision theorem 6.4,

$$H_p(K \cup CL, CL) = H_p(K, K \cap CL) = H_p(K, L)$$

for all p. The result follows from 6.9(1). □

Examples 6.11. (1) Let K be the Möbius band given in 4.15 and let L be the rim of the band. Adding a cone on L gives precisely the projective plane in 4.17(5). The relative homology groups of (K, L) are given in 4.29(3). (It is difficult to visualize $K \cup CL$ in this case because a Möbius band cannot be realized in $\mathbb{R}^3$ in such a way that the circular rim is a plane polygon with the band all on one side of it. With a possibly larger number of triangles in the Möbius band we can do the construction in $\mathbb{R}^4$; compare 2.16(5).)

(2X) Let K be connected, and let the subcomplex L have components $L_1, \ldots, L_\ell$. Then we can form a simplicial complex M by adding disjoint cones on the subcomplexes $L_1, \ldots, L_\ell$ as in the diagram below. This differs from $K \cup CL$ in that the vertices of the cones are all distinct.

Show that $\quad H_p(K, L) \cong H_p(M) \quad p > 1$
$\quad$ and $\quad H_1(K, L) \cong H_1(M) \oplus \mathbb{Z}^{\ell - 1}.$

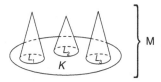

As an example of this let K be the cylinder given in 4.29(1) and L_1 and L_2 the two ends. Then M is a closed surface which is in fact a sphere, i.e. has standard form aa^{-1} (2.8). The relative homology groups of $(K, L_1 \cup L_2)$ are in 4.29(1) and the homology groups of any closed surface are given in 4.18 and 6.13 below.

(3X) Suppose that K, L and M are simplicial complexes of dimension 0, each consisting of $n \geqslant 1$ vertices. Suppose that K and L are joinable, and also that KL and M are joinable (3.26); denote the join of KL and M by KLM. (Thus when $n = 2$ the join KLM is an octahedron.) Show that $H_1(KLM) = 0$. [Hint. Take a maximal tree in the graph KL and extend it to a maximal tree in the 1-skeleton of KLM. Show, using 6.8, that the basic 1-cycles (4.9(2)) are homologous to zero.]

(4) Let S^n denote the boundary (i.e. the set of proper faces) of an $(n+1)$-simplex σ_{n+1}. Thus S^0 is two points, S^1 is a 'hollow' triangle, S^2 is a hollow tetrahedron. (In general, S^n is homeomorphic to $\{(x_1, \ldots, x_{n+1}) \in \mathbb{R}^{n+1} : x_1^2 + \cdots + x_{n+1}^2 = 1\}$ and can be called an *n-sphere*.) The homology groups of S^0 present no trouble, so suppose $n > 0$ and consider the following two subcomplexes of S^n. Let v be a vertex of σ_{n+1} and let σ_n be the n-simplex whose vertices are the remaining $n + 1$ vertices of σ_{n+1}.

Define $L_1 = $ the set of all faces of σ_n.

$L_2 = $ the cone with vertex v on the set of all proper faces of σ_n.

(The diagram shows $n = 2$.)

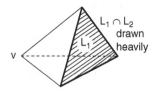

Then $S^n = L_1 \cup L_2$, $S^{n-1} = L_1 \cap L_2$. By the excision theorem,

$$H_p(S^n, L_1) \cong H_p(L_2, S^{n-1})$$

$$\cong \begin{cases} H_{p-1}(S^{n-1}) & p > 1 \\ \widetilde{H}_0(S^{n-1}) & p = 1 \end{cases}$$

by 6.9(2). Also $\widetilde{H}_0(S^{n-1}) = 0$ $(n > 1)$; $\cong \mathbb{Z}$ $(n = 1)$. Using 6.9(1) and the fact that L_1 is a cone, it follows from the above results that

$$H_p(S^n) \cong H_{p-1}(S^{n-1}) \quad p > 1$$

$$H_1(S^n) \cong \begin{cases} 0 & n > 1 \\ \mathbb{Z} & n = 1. \end{cases}$$

The second of these results, together with $H_p(S^1) = 0$ $(p > 1)$, fills in the two lines in the table below; the remaining entries follow from the first result. Namely, we have: for $n \geqslant 1$,

$$H_p(S^n) \cong \begin{cases} 0 & 0 < p < n \\ \mathbb{Z} & p = 0 \text{ or } n. \end{cases}$$

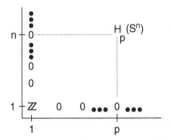

An n-cycle whose homology class generates $H_n(S^n)$ contains all the n-simplexes of S^n, oriented 'coherently', i.e. in such a way that their boundaries add up to zero. For example, S^2 is a closed surface, and such an n-cycle is a fundamental cycle for the surface (4.18).

Homology groups of closed surfaces

We shall now use the method of collapsing to calculate the first homology groups of all the closed surfaces found in Chapter 2. This cannot be done directly, since a closed surface has no principal simplex with a free face. However we can remove a 2-simplex and, using 3.34, collapse what is left to a graph, whose first homology group is straightforward to calculate. We need to know the effect of removing a 2-simplex, and that is the content of the next Lemma.

Let M be a closed surface, and let K be obtained from M by removing a single 2-simplex s; thus K is a subcomplex of M. Let z be the 1-cycle on K going once round the boundary of s and let A be the subgroup of $H_1(K)$ generated by $\{z\}_K$. (Of course $\{z\}_M = 0$ since z bounds s.)

Lemma 6.12. (1) *Suppose that M is orientable. Then*

$$i_* : H_1(K) \to H_1(M)$$

is an isomorphism.

(2) *Suppose that M is non-orientable. Then there is a short exact sequence*

$$0 \to A \xrightarrow{\text{inclusion}} H_1(K) \xrightarrow{i_*} H_1(M) \to 0.$$

It follows that $H_1(M) \cong H_1(K)/A$ and generators for $H_1(M)$ can be obtained by taking the images under i_* of generators for $H_1(K)$.

Proof Consider the homology sequence of the pair (M, K):

$$H_2(M) \xrightarrow{j_*} H_2(M, K) \xrightarrow{\partial_*} H_1(K) \xrightarrow{i_*} H_1(M) \to H_1(M, K)$$

$$\| \qquad\qquad\qquad \|$$

$$H_2(\bar{s}, \dot{s}) \cong \mathbb{Z} \qquad H_1(\bar{s}, \dot{s}) = 0$$

Here $\bar{s}$ and $\dot{s}$ are respectively the set of faces of s and the set of proper faces – the boundary – of s. The equalities follow by excision, and by 4.23(4), which gives the homology groups of $(\bar{s}, \dot{s})$.

(1) Suppose that M is orientable. Let σ be the oriented 2-simplex of M corresponding to s. Now M has a coherent orientation in which s receives the orientation of σ; also $H_2(M) \cong \mathbb{Z}$, with generator the sum of all coherently oriented 2-simplexes of M (4.18). Since j_* suppresses all 2-simplexes of M but σ, it follows that j_* will take the generator of $H_2(M)$ to $\{\sigma\}$, the homology class of σ in $H_2(M, K) = H_2(\bar{s}, \dot{s})$. However $\{\sigma\}$ generates $H_2(\bar{s}, \dot{s})$ so j_* is an isomorphism. Hence $\partial_* = 0$ and i_* is an isomorphism by exactness.

(2) Suppose that M is non-orientable, Then $H_2(M) = 0$ (4.18). As in (1), $H_2(M, K) \cong \mathbb{Z}$, with generator $\{\sigma\}$. Now $\partial_*\{\sigma\} = \{\partial\sigma\}_K = \{\pm z\}$, i.e. Im $\partial_* = A$. This proves the result, since Im $\partial_* = \operatorname{Ker} i_*$. □

Now consider a closed surface M represented in the standard form of 2.8. According to 3.34, removing a 2-simplex produces a simplicial complex K which collapses to the graph G in M given by the boundary of the polygonal region. Note that this collapse really does occur in K since the only identifications to be made in passing from the polygonal representation to the closed surface are identifications round the boundary polygon, and this is not touched during the collapse. It would not be valid, for instance, to invoke 3.32 to prove that M collapses to a point, since the collapsing of a disk starts from the boundary polygon, and the edges of M around the polygon are not in reality free.

We must now identify the graph G. Let us consider the case 2.8(2) for definiteness. Then the symbol is $a_1 a_1 \ldots a_k a_k$, as in the diagram.

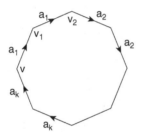

Each a_i will contain several vertices and edges of M, but the $2k$ vertices of the polygon at the endpoints of the a_i are all the same vertex v of M. (For $v_1 = v$ because they are both at the beginning of a_1, $v_2 = v_1$ because they are both at the end of a_1, and so on.) Thus each a_i gives a loop on M and all these loops have the vertex v of M in common. The case 2.8(3) is similar, and 2.8(1), the sphere aa^{-1}, differs only in that the end-points of a are distinct vertices of M. Thus the graph G to which K collapses is as follows:

(1) a simple polygonal arc in the case of the sphere aa^{-1};
(2) k loops, given by $a_1, \ldots, a_k$, with a single vertex in common, in the case of the non-orientable closed surface $a_1 a_1 \ldots a_k a_k$;
(3) $2h$ loops, given by $a_1, b_1, \ldots, a_h, b_h$, with a single vertex in common, in the case of the orientable closed surface $a_1 b_1 a_1^{-1} b_1^{-1} \ldots a_h b_h a_h^{-1} b_h^{-1}$. (See 3.35(1) for the case $h = 1$.)

Since K collapses to G, the map $i_* : H_1(G) \to H_1(K)$ is an isomorphism by 5.6. The group $H_1(G)$ is isomorphic to $\mathbb{Z}^{\mu(G)}$ and has generators given by the

152   Some general theorems

basic loops on G (see 4.9(3)). Thus

$$H_1(K) \cong \begin{cases} 0 \text{ in Case 1} \\ \mathbb{Z}^k, \text{ and is generated by } \{a_1\},\ldots,\{a_k\} \text{ in Case 2} \\ \mathbb{Z}^{2h}, \text{ and is generated by } \{a_1\},\{b_1\},\ldots,\{a_h\},\{b_h\}, \text{ in Case 3} \end{cases}$$

When M is non-orientable (Case 2), we need to know the homology class on K represented by the 1-cycle going round the boundary of a triangle s in M. Then we can apply 6.12(2). The inside of the polygon is broken up into triangles, and by adding up the boundaries of all the triangles except s, oriented clockwise in the picture, we have

$$a_1 + a_1 + a_2 + a_2 + \cdots + a_k + a_k \sim z \text{ in } K.$$

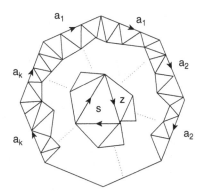

Thus the group A of 6.12(2) is generated by $2\{a_1\} + \cdots + 2\{a_k\} \in H_1(K)$. (There is little difficulty proving that the triangles of a triangulation of a disk can be oriented coherently in this way: an induction similar to that in 3.32 will work. Of course this does not coherently orient M – an impossibility – since the orientations are not coherent across the edges of M making up any of the a_i.)

Theorem 6.13. *Let M be a closed surface. Then $H_1(M)$ is as follows.*

(1) $H_1(M) = 0$ when M has standard form aa^{-1}.
(2) $H_1(M) \cong \mathbb{Z}^{k-1} \oplus \mathbb{Z}_2$ for non-orientable closed surface $a_1a_1 \ldots a_ka_k$. The classes $\{a_1\},\ldots,\{a_k\}$ on M generate $H_1(M)$ and the unique element of order two is $\{a_1\} + \cdots + \{a_k\}$.
(3) $H_1(M) \cong \mathbb{Z}^{2h}$ for the orientable closed surface $a_1b_1a_1^{-1}b_1^{-1} \ldots a_hb_ha_h^{-1}b_h^{-1}$ The classes $\{a_1\},\{b_1\},\ldots,\{a_h\},\{b_h\}$ freely generate $H_1(M)$.

Proof The cases (1) and (3) follow at once from 6.12(1) and the group $H_1(K)$, given above. Now for the case (2). The exact sequence of 6.12(2) shows that

Homology groups of closed surfaces

$H_1(M)$ is generated by the images under i_* of the generators of $H_1(K)$, i.e. by $i_*\{a_j\}_K = \{a_j\}_M$ for $j = 1, \ldots, k$. Write $\alpha_j = \{a_j\}_K$. Now $H_1(M) \cong H_1(K)/A$ and this is the quotient group of the free abelian group generated by $\alpha_1, \ldots, \alpha_k$, by the subgroup generated by 2α, where $\alpha = \alpha_1 + \cdots + \alpha_k$. Changing generators for $H_1(K)$ to $\alpha_1, \ldots, \alpha_{k-1}, \alpha$, the quotient has presentation

$$\left(\begin{array}{c} \alpha_1, \ldots, \alpha_{k-1}, \alpha \\ 2\alpha = 0 \end{array} \right.$$

and is therefore isomorphic to $\mathbb{Z}^{k-1} \oplus \mathbb{Z}_2$. The unique element of order two is the coset $\alpha + A$ corresponding to α, so the unique element of order two in $H_1(M)$ is

$$i_*\alpha = \{a_1 + \cdots + a_k\}_M = \{a_1\}_M + \cdots + \{a_k\}_M.$$

□

Thus two closed surfaces M_1 and M_2 have the same standard form (i.e. are equivalent) if and only if $H_1(M_1) \cong H_1(M_2)$: the first homology group is a sufficiently delicate invariant to distinguish the closed surfaces from one another. (In contrast, it does not distinguish a cylinder from a Möbius band, nor either of these from a simple closed polygon!)

Examples 6.14. (1X) If M is a sphere with one crosscap (projective plane) then $H_1(M) \cong \mathbb{Z}_2$. This agrees with 4.17(5). For M a sphere with two crosscaps (Klein bottle), $H_1(M) \cong \mathbb{Z} \oplus \mathbb{Z}_2$. The usual representation of a Klein bottle is shown below on the left (compare 2.2(4)) and the standard one on the right. Cutting along the diagonal b in the right-hand diagram and sticking together the two a_1's produces the left-hand picture with $c = a_2$ and $d = b$. Also $b \sim a_1 + a_2$ (add up the boundaries of triangles in the 'top half' of the right-hand picture), so by 6.13(2) $H_1(M)$ is generated by $\{c\}$ and $\{d\}$, and $\{d\}$ is the unique element of order two. To see that $d + d \sim 0$, note that $d \sim d'$ from the left half of the left-hand picture and $d' \sim -d$ from the right half. (Compare 4.29(3).)

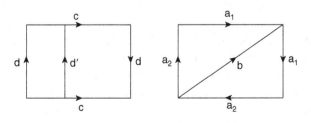

Let z be the 1-cycle going across the left-hand diagram between the midpoints of d, parallel to c. What is $\{z\}$ in terms of $\{c\}$ and $\{d\}$? What is $2\{z\}$? (In particular $2\{z\} \neq 0$.)

(2) Taking $h = 2$ in 6.13(3), possible positions for a_1, b_1, a_2, b_2 on a double torus are shown in the diagram on the left. These 1-cycles are homologous to the four shown in the diagram on the right. These four accordingly represent free generators for $H_1(M)$.

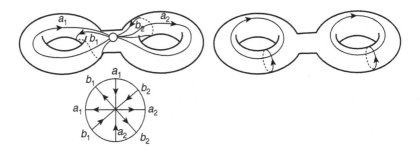

Similarly the 1-cycles drawn below, for arbitrary h, are homologous to $a_1, b_1, \ldots, a_h, b_h$ and so represent free generators for $H_1(M)$.

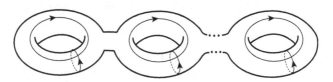

(3X) The method used to prove 6.13 can of course be used to calculate the first homology groups of a closed surface given in any polygonal representation, not just the standard one. The reader may care to try the examples of 2.16(1)–(4) directly from the representations given there.

The Euler characteristic

In this section we shall establish a connexion between the Euler characteristic of a simplicial complex (i.e. the alternating sum of the numbers of p-simplexes) and the homology groups of the simplicial complex. In particular this implies using the invariance theorem 5.13, that the Euler characteristic is a topological invariant of a simplicial complex: it is unaffected by homeomorphisms. It is interesting to note that though the Euler characteristic can be defined without mentioning homology groups, the only way (so far as I know) of proving its topological invariance is by the use of homology theory. The invariance theorem says that two simplicial complexes with different Euler characteristics cannot have homeomorphic underlying spaces. The converse it not true, even for closed surfaces, for hT and $2hP$ have the same characteristic (namely $2 - 2h$) for any $h > 0$.

Definition 6.15. Let K be a simplicial complex of dimension n and L and a subcomplex of K. Let $\alpha_p(K)$ and $\alpha_p(K, L)$ denote the numbers of p-simplexes in K and in $K \backslash L$ respectively. Then

$$\chi(K) = \sum_{p=0}^{n}(-1)^p \alpha_p(K) \quad \text{and} \quad \chi(K, L) = \sum_{p=0}^{n}(-1)^p \alpha_p(K, L)$$

are called the *Euler (or Euler–Poincaré) characteristics* of K and of K mod L respectively. (Compare 2.11.)

The connexion between these characteristics and the homology groups of K and of (K, L) is given in the next theorem, in which the K of 6.15 is assumed oriented. Let

$$\beta_p(K) = \operatorname{rank} H_p(K) \quad \text{and} \quad \beta_p(K, L) = \operatorname{rank} H_p(K, L).$$

(See A.29 for the definition of rank of a finitely generated abelian group.) These numbers are called the pth *Betti numbers* of K and of K mod L respectively. (Betti numbers were so called by H. Poincaré in honour of the Italian mathematician Enrico Betti.)

Theorem 6.16.

$$\chi(K) = \sum_{p=0}^{n}(-1)^p \beta_p(K); \quad \chi(K, L) = \sum (-1)^p \beta_p(K, L).$$

Proof The result is a purely algebraic one. Given any sequence of finitely generated abelian groups

$$0 \xrightarrow{d_{n+1}} C_n \xrightarrow{d_n} C_{n-1} \xrightarrow{d_{n-1}} \cdots \to C_p \xrightarrow{d_p} \cdots \to C_1 \xrightarrow{d_1} C_0 \xrightarrow{d_0} 0$$

where $d_{p-1} \circ d_p = 0$ for $p = 1, \ldots, n$, we can define $H_p = \operatorname{Ker} d_p / \operatorname{Im} d_{p+1}$ ($p = 0, \ldots, n$). (Compare 4.36.) Writing $\alpha_p = \operatorname{rank} C_p$, $\beta_p = \operatorname{rank} H_p$, we shall prove $\sum_{p=0}^{n}(-1)^p \alpha_p = \sum_{p=0}^{n}(-1)^p \beta_p$. The results required are then immediate.

We have short exact sequences

$$0 \to \operatorname{Ker} d_p \to C_p \xrightarrow{d_p} \operatorname{Im} d_p \to 0$$

and $\quad 0 \to \operatorname{Im} d_{p+1} \to \operatorname{Ker} d_p \to H_p \to 0.$

Thus by A.31, $\quad \alpha_p = \operatorname{rank} \operatorname{Ker} d_p + \operatorname{rank} \operatorname{Im} d_p,$

and $\quad \beta_p = \operatorname{rank} \operatorname{Ker} d_p - \operatorname{rank} \operatorname{Im} d_{p+1}$

so that $\alpha_p - \beta_p = \operatorname{rank} \operatorname{Im} d_p + \operatorname{rank} \operatorname{Im} d_{p+1}$ and $\sum_{p=0}^{n}(-1)^p(\alpha_p - \beta_p) = 0$ since d_0 and d_{n+1} are both trivial. □

Notice that A.31(3) is the special case of this result which arises when the long sequence is exact, in which case $H_p = 0$ for all p.

Examples 6.17. (1) From the definitions of chain groups it follows that, in the notation of 6.15, $\chi(K) = \chi(L) + \chi(K,L)$. This says something not quite so obvious about homology groups in view of 6.16.

Similarly if L_1 and L_2 are subcomplexes of a simplicial complex K then clearly

$$\chi(L_1 \cup L_2) = \chi(L_1) + \chi(L_2) - \chi(L_1 \cap L_2).$$

(2X) Using 6.16 and the calculated homology groups of closed surfaces in 4.18 and 6.13 we can check the values of their Euler characteristics given in 2.14.

(3X) In the notation of 6.11(3), we can calculate the Euler characteristic of KLM directly. In fact (compare 3.27(5))

$$\alpha_0(KL) = 2n; \quad \alpha_1(KL) = n^2$$
$$\alpha_0(KLM) = 3n; \quad \alpha_1(KLM) = n^2 + 2n^2 = 3n^2; \quad \alpha_2(KLM) = n^3.$$

Hence $\chi(KLM) = 3n - 3n^2 + n^3$. Since $H_0(KLM) \cong \mathbb{Z}$ and $H_1(KLM) = 0$, it follows from 6.16 that rank $H_2(KLM) = (n-1)^3$. However $H_2(KLM)$ is free abelian since KLM is two-dimensional (compare 4.9(3)), so that $H_2(KLM) \cong \mathbb{Z}^{(n-1)^3}$.

(4X) Let s be an n-simplex and let $K = \bar{s}$. Denote as usual by K^q the q-skeleton of K. Now from 6.9(3), $H_p(K) = 0$ if $0 < p < n$; consequently

(4.9(1)) $H_p(K^q) = 0$ if $0 < p < q \leqslant n$. Also $\alpha_p(K) = \binom{n+1}{p+1}$ since every choice of $p+1$ vertices from among the vertices of s spans a p-simplex of K. In an analogous way to (3) above, show that $H_q(K^q)$ is a free abelian group of rank $\binom{n}{q+1}$ if $0 < q < n$. In particular putting $q = n - 1$ we get an alternative proof of 6.11(4): $H_{n-1}(S^{n-1}) \cong \mathbb{Z}$ for all $n > 1$.

7
Two more general theorems

The theorems proved in this chapter are separated from those of Chapter 6 on account of their slightly more technical nature and the fact that we do not use them in an essential way in what follows. Both theorems assert the exactness of certain sequences. The first attempts to answer the question: given the homology groups of two subcomplexes L_1, L_2 of a simplicial complex K (not necessarily disjoint), what are the homology groups of $L_1 \cup L_2$? They obviously depend on the homology groups of $L_1 \cap L_2$ as well, and moreover on the way in which L_1 and L_2 are stuck together – for example two cylinders can be stuck together to give either a torus or a Klein bottle (see 7.5). The result is not an explicit formula for $H_p(L_1 \cup L_2)$ but an exact sequence which, with luck, will give a good deal of information. An example where it does not give quite enough information to determine a homology group of $L_1 \cup L_2$ is mentioned in 7.6(3).

The other theorem proved in this chapter, the exactness of the 'homology sequence of a triple', is a generalization of the homology sequence of a pair (see 6.1).

The Mayer–Vietoris sequence

Let K be an oriented simplicial complex of dimension n, and let L_1, L_2 be subcomplexes of K with $L_1 \cup L_2 = K$. Write $L_1 \cap L_2 = L$ and, as usual, regard $C_p(L), C_p(L_1)$ and $C_p(L_2)$ as subgroups of $C_p(K)$, for each p.

Definitions 7.1. (1) Define a homomorphism

$$\phi : H_p(L) \to H_p(L_1) \oplus H_p(L_2)$$

by

$$\{z\}_L \to (\{z\}_{L_1}, -\{z\}_{L_2}).$$

(2) Define a homomorphism

$$\psi : H_p(L_1) \oplus H_p(L_2) \to H_p(K)$$

by

$$(\{z^1\}_{L_1}, \{z^2\}_{L_2}) \to \{z^1 + z^2\}_K.$$

(3) Suppose that $z \in Z_p(K)$, and that $z = c^1 + c^2$ where $c^1 \in C_p(L_1)$ and $c^2 \in C_p(L_2)$. Then $0 = \partial z = \partial c^1 + \partial c^2$, so that $\partial c^1 = -\partial c^2$ and both ∂c^1 and ∂c^2 must belong to $C_{p-1}(L)$ – indeed to $Z_{p-1}(L)$ since $\partial \circ \partial = 0$.

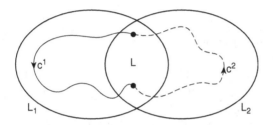

Define a homomorphism

$$\Delta : H_p(K) \to H_{p-1}(L)$$

by

$$\{z\}_K \to \{\partial c^1\}_L.$$

To prove this makes sense, suppose that $\{w\}_K = \{z\}_K$, and that $w = c^3 + c^4$ where $c^3 \in C_p(L_1)$ and $c^4 \in C_p(L_2)$. We have to show $\{\partial c^1\}_L = \{\partial c^3\}_L$. Now $z - w = \partial c$ for some $c \in C_{p+1}(K)$, so that writing $c = c^5 + c^6$ ($c^5 \in C_{p+1}(L_1)$, $c^6 \in C_{p+1}(L_2)$), we have $c^1 - c^3 - \partial c^5 = -(c^2 - c^4 - \partial c^6)$. Accordingly both sides of this equation belong to $C_p(L)$, whence

$$\partial(c^1 - c^3 - \partial c^5) = \partial c^1 - \partial c^3 \in B_{p-1}(L),$$

i.e. $\{\partial c^1\}_L = \{\partial c^3\}_L$ as required.

The verifications that (1) and (2) also make sense are left to the reader. We now fit the three homomorphisms ϕ, ψ, Δ into an exact sequence, called the *Mayer–Vietoris sequence* of the triad $(K; L_1, L_2)$.

Theorem 7.2. *The following sequence is exact.*

$$0 \to H_n(L) \xrightarrow{\phi} H_n(L_1) \oplus H_n(L_2) \xrightarrow{\psi} H_n(K) \xrightarrow{\Delta} H_{n-1}(L)$$
$$\xrightarrow{\phi} H_{n-1}(L_1) \oplus H_{n-1}(L_2) \xrightarrow{\psi} \ldots \xrightarrow{\Delta} H_0(L) \xrightarrow{\phi} H_0(L_1) \oplus H_0(L_2)$$
$$\xrightarrow{\psi} H_0(K) \to 0.$$

Proof (1) Exactness at $H_p(L)$. Let $z \in Z_{p+1}(K)$ and $z = c^1 + c^2$ where $c^1 \in C_{p+1}(L_1)$, $c^2 \in C_{p+1}(L_2)$. Then

$$\phi\Delta\{z\} = (\{\partial c^1\}_{L_1}, -\{\partial c^1\}_{L_2}) = (\{\partial c^1\}_{L_1}, \{\partial c^2\}_{L_2}) = (0,0).$$

Hence Im $\Delta \subset$ Ker ϕ.

Conversely, suppose that $u \in Z_p(L)$ satisfies $\phi\{u\} = (0,0)$, i.e. $\{u\}_{L_1} = 0$ and $\{u\}_{L_2} = 0$. Thus $u = \partial x^1 = \partial x^2$, say, so $\partial(x^1 - x^2) = 0$, i.e. $x^1 - x^2 \in Z_p(K)$. Furthermore $\Delta\{x^1 - x^2\}_K = \{\partial x^1\}_L = \{u\}_L$. Hence Ker $\phi \subset$ Im Δ.

(2) Exactness at $H_p(L_1) \oplus H_p(L_2)$. Let $z \in Z_p(L)$; then $\psi\phi\{z\} = \{z-z\}_L = 0$. Hence Im $\phi \subset$ Ker ψ.

Conversely, suppose that $(\{u^1\}, \{u^2\}) \in H_p(L_1) \oplus H_p(L_2)$ lies in Ker ψ, i.e. that $\{u^1 + u^2\}_K = 0$. Thus $u^1 + u^2 = \partial x$ for some $x \in C_{p+1}(K)$. Let $\rho_\alpha : C_{p+1}(K) \to C_{p+1}(L_\alpha), \alpha = 1, 2$ be the restriction homomorphisms which change to zero the coefficient of any $(p+1)$-simplex not in L_α; I claim that $u^1 - \partial\rho_1 x \in Z_p(L)$ and that $\phi\{u^1 - \partial\rho_1 x\} = (\{u^1\}, \{u^2\})$. To see this, note that

$$x - \rho_1 x - \rho_2 x \in C_{p+1}(L)$$

so that

$$\partial x - \partial\rho_1 x - \partial\rho_2 x \in B_p(L)$$

i.e.

$$(u^1 - \partial\rho_1 x) + (u^2 - \partial\rho_2 x) \in B_p(L).$$

But the first bracket belongs to $C_p(L_1)$ and the second to $C_p(L_2)$; hence both belong to $C_p(L)$ and indeed since u^1 and u^2 are cycles both brackets belong to $Z_p(L)$. Finally $u^\alpha - \partial\rho_\alpha x \sim u^\alpha$ on $L_\alpha (\alpha = 1, 2)$ and $-(u^1 - \partial\rho_1 x) \sim u^2 - \partial\rho_2 x$ on L, and hence on L_2. Applying the definition of ϕ the required result follows. Thus Ker $\psi \subset$ Im ϕ. (The reader may like to draw a diagram to illustrate this inclusion, say in the case $p = 1$.)

(3) Exactness at $H_p(K)$. Let $(\{z^1\}, \{z^2\}) \in H_p(L_1) \oplus H_p(L_2)$. Then $\Delta\psi(\{z^1\}, \{z^2\}) = \{\partial z^1\}_L = 0$ since z^1 is a cycle. Hence Im $\psi \subset$ Ker Δ.

Conversely, suppose that $u \in Z_p(K)$ satisfies $\Delta\{u\} = 0$, i.e. writing $u = c^1 + c^2$ in the usual way, that $\{\partial c^1\}_L = 0$. This means that $\partial c^1 = \partial x$ for some $x \in C_p(L)$. Then $c^1 - x \in Z_p(L_1)$ and $c^2 + x \in Z_p(L_2)$ since u is a cycle. Since

$$\psi(\{c^1 - x\}_{L_1}, c^2 + x\}_{L_2}) = \{c^1 + c^2\}_K = \{u\}_K$$

this shows that Ker $\Delta \subset$ Im ψ. □

Remark 7.3. There is a reduced form of the Mayer–Vietoris sequence in which zero-dimensional homology groups are replaced by reduced groups (4.8) and Δ, ϕ and ψ are suitably altered at the lower end of the sequence. The resulting sequence is still exact.

Examples 7.4. (1) If L is empty, then K is the union of disjoint subcomplexes L_1, L_2 and from the Mayer–Vietoris sequence

$$H_p(K) \cong H_p(L_1) \oplus H_p(L_2)$$

for all p. This result extends in an obvious way to the union of any number of disjoint subcomplexes, and it is not hard to give a direct proof.

(2X) Suppose that L is connected. Then

$$\phi : H_0(L) \to H_0(L_1) \oplus H_0(L_2)$$

is monic (injective). (If L is empty there is nothing to prove; otherwise let v be a vertex of L so that $\{v\}$ generates the infinite cyclic group $H_0(L)$. Now

$$\phi\{v\} = (\{v\}_{L_1}, -\{v\}_{L_2}) \neq (0, 0)$$

and since $H_0(L_1) \oplus H_0(L_2)$ has no non-zero element of finite order this is the result.)

(3X) Suppose that L consists of a single point. Then $H_p(L) = 0$ for $p > 0$ so that from the Mayer–Vietoris sequence $H_p(K) \cong H_p(L_1) \oplus H_p(L_2)$ for $p > 1$. From (2) it follows that this holds for $p = 1$ as well. What about $p = 0$?

(4X) Suppose that L consists of two points, v and w say. Show that

$$\phi : H_0(L) \to H_0(L_1) \oplus H_0(L_2)$$

has infinite cyclic kernel generated by $\{v - w\}$ if v and w lie in the same component of L_1 and in the same component of L_2, and is monic (injective) otherwise. (Note that if v and w lie in different components of L_1 then $\lambda v + \mu w \sim 0$ on L_1 implies $\lambda = \mu = 0$. Compare the proof of 4.11.) Hence show that

$$H_p(K) \cong H_p(L_1) \oplus H_p(L_2), \quad p > 1$$
$$H_1(K) \cong H_1(L_1) \oplus H_1(L_2) \oplus A$$

where $A = 0$ or $A \cong \mathbb{Z}$. In particular $A \cong \mathbb{Z}$ when K is connected, and the diagram shows where the extra 1-cycle comes from. Find examples to illustrate the case $A = 0$.

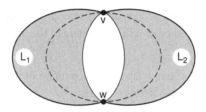

(5X) Some earlier results can be reproved using the Mayer–Vietoris sequence, for instance 6.11(4) (homology groups of S^n), 6.6 (collapsing) and 6.12 (effect of removing a 2-simplex from a closed surface). The reader is invited to try these examples, and to look for others in Chapter 6 amenable to this alternative treatment.

(6X) Let K be an oriented simplicial complex in $\mathbb{R}^N$, entirely contained in the subspace $x_1 = 0$ (where $(x_1, \ldots, x_N)$ are coordinates in $\mathbb{R}^N$). Let C_1 be the cone on K with vertex $(1, 0, \ldots, 0)$ and C_2 the cone on K with vertex $(-1, 0, \ldots, 0)$, so that $C_1 \cap C_2 = K$. We call $C_1 \cup C_2$ a *suspension* of K and denote it by SK. Show that

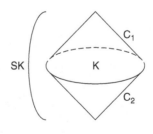

$$H_p(SK) \cong H_{p-1}(K) \quad (p > 1)$$
$$H_1(SK) \cong \widetilde{H}_0(K).$$

Which closed surfaces, if any, are suspensions of graphs?

The remaining examples will be of a more informal nature than those given above, in that we shall not worry very much about the triangulations of the objects in question but rather witness the Mayer–Vietoris sequence in action by sticking together simple objects to make more complicated ones. In this way I hope that the geometrical content of the sequence will become clearer.

Example 7.5. If we identify the ends of two cylinders, the result can be either a torus or a Klein bottle.

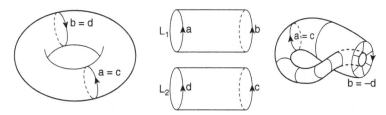

We shall assume the following: any triangulation L_1 of a cylinder has $H_1(L_1) \cong \mathbb{Z}$, with generator a cycle going round one end; $H_2(L_1) = 0$. (Compare 4.17(1); the result could indeed be proved by a collapsing argument.) Now let us see whether the Mayer–Vietoris sequence enables us to calculate the homology groups of a torus and a Klein bottle, which are already known from 6.13. Note that in either case $L_1 \cap L_2 = L$ consists of two disjoint circles.

$$H_2(L_1) \oplus H_2(L_2) \xrightarrow{\psi_2} H_2(K) \xrightarrow{\Delta_2} H_1(L) \xrightarrow{\phi_1} H_1(L_1) \oplus H_1(L_2) \xrightarrow{\psi_1}$$
$$\; 0 0 \mathbb{Z} \oplus \mathbb{Z} \mathbb{Z} \;\oplus\; \mathbb{Z}$$

$$\xrightarrow{\psi_1} H_1(K) \xrightarrow{\Delta_1} H_0(L) \xrightarrow{\phi_0} H_0(L_1) \oplus H_0(L_2) \xrightarrow{\psi_0} H_0(K) \to 0.$$
$$ \mathbb{Z} \oplus \mathbb{Z} \mathbb{Z} \;\oplus\; \mathbb{Z}$$

Consider

$$\phi_1 : H_1(L) \to H_1(L_1) \oplus H_1(L_2).$$

The groups are

$$\mathbb{Z} \oplus \mathbb{Z} \mathbb{Z} \;\oplus\; \mathbb{Z}$$

with generators

$$\{a\}, \{b\}_L (\{a\}_{L_1}, 0), (0, \{c\}_{L_2})$$
$$\phantom{\{a\}, \{b\}_L} = \alpha_1, \alpha_2 \text{say}.$$

Now

$$\phi_1\{a\}_L = (\{a\}_{L_1}, -\{a\}_{L_2}) = (\{a\}_{L_1}, -\{c\}_{L_2}) = \alpha_1 - \alpha_2$$
$$\phi_1\{b\}_L = (\{b\}_{L_1}, -\{b\}_{L_2}) = (\{a\}_{L_1}, \pm\{c\}_{L_2}) = \alpha_1 \pm \alpha_2$$

where the sign is $+$ or $-$ for the Klein bottle or torus respectively.
From this it follows that

$$H_2(K) \cong \operatorname{Ker} \phi_1 \cong \begin{cases} \mathbb{Z} & \text{generator } \{a - b\}_L \text{ for torus} \\ 0 & \text{for Klein bottle.} \end{cases}$$

A generator for $H_2(K)$ in the case of the torus is any element going to $\{a-b\}_L$, under Δ_2. Such an element is the homology class of the sum of all 2-simplexes of K, coherently oriented.

A similar examination of ϕ_0 reveals that $\operatorname{Im} \Delta_1 = \operatorname{Ker} \phi_0 \cong \mathbb{Z}$, with generator $\{v^1 - v^2\}_L$ where v^1, v^2 are vertices in the two components of L.

Now there is a short exact sequence

$$0 \to \operatorname{Im} \psi_1 \overset{\text{inclusion}}{\to} H_1(K) \overset{\Delta_1}{\to} \operatorname{Im} \Delta_1 \to 0.$$

Since $\operatorname{Im} \Delta_1 \cong \mathbb{Z}$, this sequence splits (A.19). Also
$\operatorname{Im} \psi_1 \cong (H_1(L_1) \oplus H_1(L_2))/\operatorname{Im} \phi_1$ by the homomorphism theorem (A.15).
This latter group has presentation

$$\begin{pmatrix} \alpha_1, \alpha_2 \\ \alpha_1 - \alpha_2 \\ \alpha_1 \pm \alpha_2 \end{pmatrix} (+ \text{ for Klein bottle, } - \text{ for torus})$$

and so is isomorphic to $\mathbb{Z}_2$ (generator $\alpha_1 + \operatorname{Im} \phi_1$) or $\mathbb{Z}$ (generator $\alpha_1 + \operatorname{Im} \phi_1$) according as the sign is $+$ or $-$. A generator for $\operatorname{Im} \psi_1$ is therefore $\psi_1(\alpha_1)$ in either case, i.e. $\{a\}_K$. An element of $H_1(K)$ going under Δ_1 to $\{v^1 - v^2\}_L$ is $\{z\}$, where z is indicated heavily on the diagrams below.

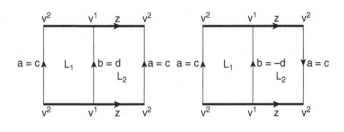

Thus $H_1(\text{torus}) \cong \mathbb{Z} \oplus \mathbb{Z}$; $H_1(\text{Klein bottle}) \cong \mathbb{Z} \oplus \mathbb{Z}_2$ with generators $\{z\}$ and $\{a\}$ in each case.

Examples 7.6. (1X) In the spirit of 7.5, split a Klein bottle into two Möbius bands and calculate the homology groups of the Klein bottle from those of the Möbius bands. L is a single circle, drawn heavily.

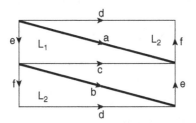

Assume that $H_1(L_1) \cong \mathbb{Z}$ (generator $\{c\}$); $H_1(L_2) \cong \mathbb{Z}$ (generator $\{d\}$); $H_1(L) \cong \mathbb{Z}$ (generator $\{a+b\}$); $a + b \sim 2c$ on L_1; $a + b \sim 2d$ on L_2.

(2X) Let M_1 and M_2 be two orientable closed surfaces and let L_1, L_2 be obtained by removing a 2-simplex from, respectively, M_1 and M_2. Then $M = L_1 \cup L_2$ is a connected sum $M_1 \# M_2$ (2.9). Use the Mayer–Vietoris sequence to show that

$$H_1(M) \cong H_1(M_1) \oplus H_1(M_2).$$

[Hints. Show that $H_1(L_1) \cong H_1(M_1)$, either by quoting 6.12 or by using the Mayer–Vietoris sequence. Then use the sequence again to show $H_1(M) \cong H_1(L_1) \oplus H_1(L_2)$.] Thus once the first homology group of any torus is known this reproduces 6.13(2), at least up to isomorphism.

(3X) The reason why (2) works is that the boundary of the 'missing' 2-simplex is homologous to zero on L_1, and on L_2. If we stick together a torus and a projective plane, say, this is no longer the case. Make the following *assumption*: $L_2 = $ triangulation of a projective plane, minus a 2-simplex, has $H_1(L_2) \cong \mathbb{Z}$ and the boundary of the 'missing' 2-simplex represents twice a generator. Now use the Mayer–Vietoris sequence to calculate the first homology group of the connected sum in question. The above assumption follows, of course, from the collapsing arguments of Chapter 6 – but then so does the final result! Removing a 2-simplex from a triangulation of a projective plane does in fact give a triangulation of a Möbius band, so the assumption is related to a calculation in 4.15.

It is worth noting that the Mayer–Vietoris sequence cannot by itself be used to prove the assumption, nor even that $H_1(L_2) \cong \mathbb{Z}$ given $H_1(P) = \mathbb{Z}_2$ and $H_2(P) = 0$. For the sequence only provides us with a short exact sequence

$$0 \to \mathbb{Z} \to H_1(L_2) \to \mathbb{Z}_2 \to 0$$

from which $H_1(L_2)$ could be isomorphic either to $\mathbb{Z}$ or to $\mathbb{Z} \oplus \mathbb{Z}_2$. Compare A.20(2).

By writing a general non-orientable closed surface kP as

$$P \# \tfrac{1}{2}(k-1)T \quad k \text{ odd}$$
$$\text{or} \quad 2P \# \tfrac{1}{2}(k-2)T \quad k \text{ even}$$

(compare the discussion following 2.10) the above result can be used to calculate $H_1(kP)$, making use of the above assumption and an analogous one for $2P =$ Klein bottle. Can the Mayer–Vietoris sequence be used to calculate $H_1(2P)$ using just the above assumption about P?

(4X) Again in the spirit of 7.5, calculate the homology groups of the following objects.

Union of a torus and a disk spanning a meridian circle. (The disk has the homology groups of a point, since (3.32) it collapses to a point.)

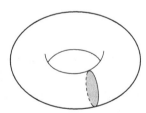

Union of a torus and two disks spanning parallel meridian circles.

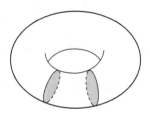

Union of a double torus and a disk spanning the 'waist'.

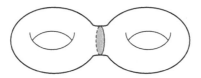

★ Homology sequence of a triple

The theorem about to be proved generalizes 6.1, the exactness of the homology sequence of a pair. It will be used in Chapter 9, but not in an essential way, and is included here for completeness. There are several ways of proving the theorem: undoubtedly the 'right' way is to use some purely algebraic machinery such as that in 4.36 et seq. to deduce the result from an exact sequence of chain complexes; or an ad hoc proof analogous to that of 6.1 can be constructed. We shall adopt a third alternative, and deduce it from 6.1 itself, together with one of the more picturesque results in the theory of exact sequences, known as the Braid Theorem (see 7.9 below).

The notation here is chosen to be close to that in Chapter 9, where the theorem will be used. Let K and N be subcomplexes of a simplicial complex M, and suppose that $K \subset N$.

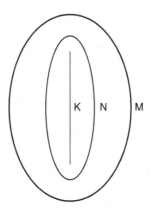

Thus $K \subset N \subset M$. There are three pairs here: (N, K), (M, K) and (M, N). The object is to fit their relative homology groups into an exact sequence. As usual, all chain groups are regarded as subgroups of the chain groups of M.

Definitions 7.7. Define $\alpha = \alpha_p : H_p(N, K) \to H_p(M, K)$ by $\alpha\{z\}_{N,K} = \{z\}_{M,K}$ for $z \in Z_p(N, K)$.

Define
$$\beta = \beta_p : H_p(M,K) \to H_p(M,N)$$
by
$$\beta\{z\}_{M,K} = \{jz\}_{M,N} \text{ for } z \in Z_p(M,K)$$
where
$$j : C_p(M) \to C_p(M,N)$$
is the usual homomorphism.
Define
$$\gamma = \gamma_p : H_p(M,N) \to H_{p-1}(N,K)$$
by
$$\gamma\{z\}_{M,N} = \{j'\partial z\}_{N,K} \quad z \in Z_p(M,N)$$
where
$$j' : C_{p-1}(N) \to C_{p-1}(N,K)$$
is the usual homomorphism.

The verifications that these are well-defined are left to the reader.

Theorem 7.8. *If M has dimension n, then the sequence*

$$0 \to H_n(N,K) \xrightarrow{\alpha} H_n(M,K) \xrightarrow{\beta} H_n(M,N) \xrightarrow{\gamma} H_{n-1}(N,K) \xrightarrow{\alpha} \cdots$$
$$\cdots \to H_1(M,N) \xrightarrow{\gamma} H_0(N,K) \xrightarrow{\alpha} H_0(M,K) \xrightarrow{\beta} H_0(M,N) \to 0$$

is exact. It is called the homology sequence of the triple (M,N,K).

Note that if $K = \emptyset$ this reduces to the homology sequence of the pair (M,N). As mentioned above the proof will depend on 6.1 and a purely algebraic result on exact sequences, which is as follows.

Homology sequence of a triple

Consider the diagram of abelian groups and homomorphisms below.

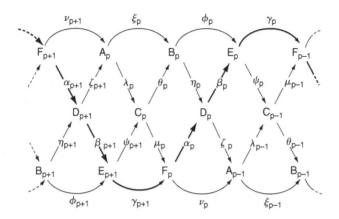

There are four 'sine-wave' sequences in the diagram, in which the homomorphisms are, respectively, labelled $\lambda, \mu, \nu; \xi, \eta, \zeta; \theta, \phi, \psi;$ and α, β, γ. Apart from $\alpha, \beta,$ and γ, the notation in this theorem has no connexion with previous usage – sheer pressure of notation has necessitated the re-use of several symbols. The 'regions' of the diagram are three- or four-sided, and to say that the diagram 'commutes' is to say that two alternative routes from one corner group of a region to another are always the same homomorphism – thus $\eta_p \circ \theta_p = \alpha_p \circ \mu_p$ for all p, etc.

7.9. Braid theorem *Suppose that the above diagram commutes and that the first three sequences mentioned above are exact. Suppose also that $\beta_p \circ \alpha_p = 0$ for all p. Then the sequence of α's, β's and γ's is exact.*

Proof The proof is of a kind called *diagram-chasing* and we shall only give about half of it, leaving the reader to pursue the rest. It is enough to prove exactness at F_p, D_p and E_p.

(1) Exactness at F_p

$$\alpha_p \circ \gamma_{p+1} = \alpha_p \circ \mu_p \circ \psi_{p+1} = \eta_p \circ \theta_p \circ \psi_{p+1} = 0$$

by exactness of the θ, ϕ, ψ sequence.

Hence Im $\gamma_{p+1} \subset \operatorname{Ker} \alpha_p$.

Conversely suppose $\alpha_p f = 0, f \in F_p$. Then $0 = \zeta_p \alpha_p f = \nu_p f$; hence $f = \mu_p c$ for some $c \in C_p$. Thus $0 = \alpha_p \mu_p c = \eta_p \theta_p c$, so $\theta_p c \in \operatorname{Ker} \eta_p = \operatorname{Im} \xi_p$. Hence $\theta_p c = \xi_p a$ for some $a \in A_p$. We now have $\theta_p(\lambda_p a) = \xi_p a = \theta_p c$; hence

$c - \lambda_p a \in \operatorname{Ker} \theta_p = \operatorname{Im} \psi_{p+1}$. This gives $c - \lambda_p a = \psi_{p+1} e$ for some $e \in E_{p+1}$, and

$$\gamma_{p+1} e = \mu_p \psi_{p+1} e = \mu_p (c - \lambda_p a) = \mu_p c = f.$$

Hence $\operatorname{Ker} \alpha_p \subset \operatorname{Im} \gamma_{p+1}$.

(2) Exactness at D_p. One half of this is given; for the other half suppose $\beta_p d = 0$ for some $d \in D_p$. Then $0 = \psi_p \beta_p d = \lambda_{p-1} \zeta_p d$. Hence $\zeta_p d \in \operatorname{Ker} \lambda_{p-1} = \operatorname{Im} \nu_p$, so $\zeta_p d = \nu_p f$ for some $f \in F_p$. Thus $\zeta_p \alpha_p f = \nu_p f = \zeta_p d$, so $d - \alpha_p f \in \operatorname{Ker} \zeta_p = \operatorname{Im} \eta_p$. This gives $d - \alpha_p f = \eta_p b$ for some $b \in B_p$, and since $\phi_p b = \beta_p \eta_p b = 0$ (since $\beta_p d = 0$), $b \in \operatorname{Ker} \phi_p = \operatorname{Im} \theta_p$, and $b = \theta_p c$ for some $c \in C_p$. Thus

$$\alpha_p(\mu_p c + f) = \eta_p \theta_p c + \alpha_p f = \eta_p b + d - \eta_p b = d.$$

Hence $\operatorname{Ker} \beta_p \subset \operatorname{Im} \alpha_p$.

The proof of exactness at E_p is left to the reader. □

To deduce that the homology sequence of a triple is exact we make the following substitutions:

$$A_p = H_p(K), \quad B_p = H_p(M), \quad C_p = H_p(N), \quad D_p = H_p(M, K),$$
$$E_p = H_p(M, N), \quad F_p = H_p(N, K).$$

The homomorphisms being the usual ones, and $\alpha_p, \beta_p, \gamma_p$ being as in 7.7, the three sequences named first in the preamble to 7.9 become the homology sequences of the pairs (N, K), (M, K) and (M, N) respectively, while the fourth sequence becomes the homology sequence of the triple (M, N, K). The details of checking that the hypotheses of the Braid Theorem are satisfied are left to the reader. □

Example 7.10. (X) In the notation of 7.8, suppose that $N \searrow K$. Using $H_p(N, K) = 0$ (see the proof of 6.6) and 7.8 deduce that $H_p(M, K) \cong H_p(M, N)$ for all p. What are the corresponding results (i) if $M \searrow N$; (ii) if $M \searrow K$? ★

8
Homology modulo 2

The homology theory developed so far applies only to oriented simplicial complexes. In this chapter we shall show how to make adjustments in the theory so that it applies to unoriented simplicial complexes. Roughly speaking if an oriented simplex σ and the same simplex with opposite orientation, $-\sigma$, are to be regarded as giving the same chain, then $-1 = +1$ and so the coefficients must be regarded as lying in the group (or field) $\mathbb{Z}_2$ instead of in the group $\mathbb{Z}$. A satisfactory theory can be developed on these lines, though it is weaker than the theory with integral coefficients: for example the closed surfaces nT and $2nP$ have the same 'modulo 2' homology groups for any $n \geq 1$, and so cannot be distinguished by means of these groups.

The point of including this theory here is that with its aid certain results in Chapter 9 on the embedding of graphs in surfaces can be proved for both orientable and non-orientable closed surfaces.

Definitions 8.1. Let K be a simplicial complex of dimension n, and let $s_p^1, \ldots, s_p^{\alpha_p}$ be the p-simplexes of K for $p = 0, \ldots, n$. A *p-chain mod 2* on K is a formal sum

$$\lambda_1 s_p^1 + \cdots + \lambda_{\alpha_p} s_p^{\alpha_p}$$

where the coefficients λ_i are elements of the field $\mathbb{Z}_2$ so that each coefficient is 0 or 1. The set of p-chains mod 2 is denoted by $C_p(K; 2)$ (or $C_p(K; \mathbb{Z}_2)$) and it is a vector space over the field $\mathbb{Z}_2$ with the definitions of *addition* and *scalar multiplication* as follows:

$$\sum \lambda_i s_p^i + \sum \lambda_i' s_p^i = \sum (\lambda_i + \lambda_i') s_p^i$$
$$\lambda \left(\sum \lambda_i s_p^i \right) = \sum \lambda \lambda_i s_p^i, \quad \lambda \in \mathbb{Z}_2.$$

For values of p other than $0, \ldots, n$ we define $C_p(K; 2) = 0$.

The *boundary homomorphism mod 2*

$$\partial = \partial_p : C_p(K; 2) \to C_{p-1}(K; 2)$$

is defined for $0 < p \leqslant n$ by

$$\partial \left(\sum \lambda_i s_p^i \right) = \sum \lambda_i \partial s_p^i,$$

where

$$\partial(v^0 \ldots v^p) = \sum_{i=0}^{p} (v^0 \ldots \widehat{v}^i \ldots v^p),$$

and is defined to be trivial otherwise. In fact ∂ is a linear map.

The *augmentation mod 2* is the linear map

$$\varepsilon : C_0(K; 2) \to \mathbb{Z}_2$$

defined by

$$\varepsilon \left(\sum \lambda_i s_0^i \right) = \sum \lambda_i.$$

Thus, in the diagram below, we have (all mod 2)

$$\partial t^1 = a + e + d$$
$$\partial t^2 = b + c + e$$
$$\partial(t^1 + t^2) = a + b + c + d + 2e$$
$$= a + b + c + d \quad \text{since } 2e = 0$$
$$\partial a = u + v, \quad \text{etc.}$$
$$\partial(a + b + c + d) = u + v + v + w + w + x + x + u$$
$$= 0.$$

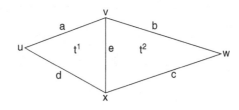

As another example, consider the diagram of a Möbius band in 4.15. Removing all arrows which indicate orientations we have

$$\partial(t^1 + t^2 + t^3 + t^4 + t^5) = e^2 + e^3 + e^4 + e^5 + e^6 \pmod{2};$$

that is 'the mod 2 boundary of the whole band is its rim'. Putting it the other way round, the rim is, mod 2, a boundary. This is false in ordinary homology: there is no 2-chain on the oriented K whose boundary is $e^2 + \cdots + e^6 \in C_1(K)$. This suggests that, for simple closed polygons, the property of enclosing a region is closely connected with being a boundary mod 2. (This will be made precise and proved in 9.19.)

An example where the Möbius band occurs as a region in something bigger is given by the Klein bottle of 2.2(4), where the shaded triangles form a Möbius band. In this case the rim of the band represents, in ordinary homology, $\pm 2\{a + b + c\}$, which is not zero (compare the last sentence of 6.14(1)). When we pass to homology mod 2 the corresponding class is zero, since the rim is, mod 2, a boundary.

Remarks 8.2. ★(1) We can form the group of p-chains with coefficients in any abelian group A in an analogous way to that given above for $A = \mathbb{Z}_2$, and this group is denoted by $C_p(K; A)$. The case $A = \mathbb{Z}$ is the one used hitherto. In the formula for ∂, a sign $\pm$ must be introduced, $+$ if i is even and $-$ if i is odd, in order that $\partial \circ \partial$ should be zero, but for $A = \mathbb{Z}_2$ the sign is redundant. If A is a field, such as the field of real numbers, rational numbers or complex numbers, then $C_p(K; A)$ is a vector space over A, using the same definitions of scalar multiplication and addition as for $A = \mathbb{Z}_2$; furthermore in that case ∂ is a linear map. ★

(2) Suppose that $f : A \to B$ is a map between vector spaces over $\mathbb{Z}_2$, and that $f(a_1 + a_2) = f(a_1) + f(a_2)$ for all a_1, a_2, in A. Then f is automatically a linear map, i.e. f also satisfies $f(\lambda a) = \lambda f(a)$ for all $\lambda \in \mathbb{Z}_2$ and $a \in A$. (There are only two cases, $\lambda = 0$ and $\lambda = 1$, to check!)

Chain complexes mod 2, augmented chain complexes mod 2, boundaries and cycles are all defined as in the case of $\mathbb{Z}$ coefficients (4.5 and 4.8). We also have the following result.

Proposition 8.3. *For any p,*

$$\partial_{p-1} \circ \partial_p : C_p(K; 2) \to C_{p-2}(K; 2)$$

is the trivial homomorphism. Furthermore

$$\varepsilon \circ \partial_1 : C_1(K;2) \to \mathbb{Z}_2$$

is trivial. □

Thus Im $\partial_{p+1} \subset$ Ker ∂_p for all p, and we can at any rate form a quotient group

$$H_p(K;2) = \text{Ker } \partial_p / \text{Im } \partial_{p+1} = Z_p(K;2)/B_p(K;2)$$

where $Z_p(K;2)$ and $B_p(K;2)$ are regarded as just abelian groups. Then $H_p(K;2)$ is called the pth *homology group of K* mod 2 (or *with $\mathbb{Z}_2$ coefficients*). In fact $H_p(K;2)$ is a vector space over $\mathbb{Z}_2$ with scalar multiplication defined by

$$\lambda(z + B_p(K;2)) = \lambda z + B_p(K;2)$$

where $z \in Z_p(K;2)$ and $\lambda \in \mathbb{Z}_2$. Thus as a vector space, and hence as an abelian group also,

$$H_p(K;2) \cong \mathbb{Z}_2 \oplus \mathbb{Z}_2 \oplus \cdots \oplus \mathbb{Z}_2.$$

The number of summands is the dimension of the vector space $H_p(K;2)$, and is called the pth *connectivity number* of K. It is denoted $\widehat{\beta}_p(K)$.

There is no analogue of 'torsion coefficients', for every vector space is free in the sense that

$$\lambda x = 0 \quad (\lambda \in \text{field}, x \in \text{vector space}) \Rightarrow \lambda = 0 \quad \text{or} \quad x = 0.$$

Of course considered just as an abelian group, $\mathbb{Z}_2 \oplus \cdots \oplus \mathbb{Z}_2$ is far from free, since $\lambda x = 0$ for any $x \in \mathbb{Z}_2 \oplus \cdots \oplus \mathbb{Z}_2$ when $\lambda = 2 \in \mathbb{Z}$.

Two elements z, z' in $Z_p(K;2)$ are called *homologous mod 2* if $z - z' \in B_p(K;2)$. We then write $z \sim z'$ (mod 2). The homology class mod 2 of $z \in Z_p(K;2)$, namely $z + B_p(K;2)$, is denoted $\{z\}$ or $\{z\}_K$.

We also define the *reduced group* (again a vector space)

$$\widetilde{H}_0(K;2) = \text{Ker } \varepsilon / \text{Im } \partial_1.$$

Examples 8.4. (1) Let M be a closed surface and let $c \in C_2(M;2)$ be the sum of all the triangles of M. Then c is a cycle, because every edge of M is on exactly two triangles of M. Notice that this does not assume M to be orientable. Using a similar argument to that in 4.18 it follows that $H_2(M;2) \cong \mathbb{Z}_2$, with c (strictly $\{c\}$) as a generator. We call c the *fundamental cycle mod 2* on M. Thus

$H_2(M; 2)$ does not distinguish between orientable and non-orientable closed surfaces. (But see 8.9(3).)

(2X) Suppose that K is connected. A basis for $Z_1(K; 2)$ is obtained in the following way, exactly analogous to that for $Z_1(K)$ in 4.9(2). Let T be a maximal tree in the 1-skeleton K^1 of K, and let $e^1, \ldots, e^\mu$ be the edges of K^1 not in T. Then for each $i = 1, \ldots, \mu$, $T + e^i$ contains a unique loop and the sum of the edges in this loop is a 1-cycle z^i mod 2. We have: $z^1, \ldots, z^\mu$ is a basis for $Z_1(K; 2)$. If K is not connected put together bases for the 1-cycles in each component of K.

In particular if K is a connected graph ($K^1 = K$) then $H_1(K; 2) \cong \mathbb{Z}_2 \oplus \cdots \oplus \mathbb{Z}_2$ (μ summands) with basis $z^1, \ldots, z^\mu$. If K has several components then $H_1(K; 2)$ is the direct sum of the $H_1(K_i; 2)$ where K_i runs through the components of K.

(3X) Let K be obtained from a closed surface M by removing a single 2-simplex s, and let z be the 1-cycle mod 2 on K given by the sum of the edges of s. Then $z \sim 0$ (mod 2) on K whether or not M is orientable – compare 7.6(3).

(4) $H_0(K; 2)$ has dimension equal to the number of components of K, with basis consisting of the homology classes mod 2 of one vertex from each component. The proof given in 4.11 of the corresponding statement about $H_0(K)$ also proves the present statement, subject only to the obvious changes from $\mathbb{Z}$ to $\mathbb{Z}_2$ in various places. Similarly, if K is non-empty,

$$H_0(K; 2) \cong \tilde{H}_0(K; 2) \oplus \mathbb{Z}_2.$$

Again the proof given before (4.13) is valid, remembering that short exact sequences of vector spaces and linear maps always split (see after A.19).

(5X) Let K be the triangulation of a Möbius band given in 4.15. The calculation of $H_1(K; 2)$ proceeds in exactly the same way as that for $H_1(K)$, except that all arrows can be removed and minus signs changed to plus signs. We can either regard the presentations as defining abelian groups in which, additionally to the relations given, every element has order two, or better, as defining vector spaces over $\mathbb{Z}_2$ as quotients of vector spaces by subspaces. In the latter case the theory is exactly parallel to that expounded in the Appendix and we can manipulate the presentations until a basis for the quotient vector space emerges. Of course these manipulations are enormously simplified by the fact that $1 + 1 = 0$, and at the end of the calculation there will be generators but no relations in the presentation.

The result for the Möbius band is

$$H_1(K; 2) \cong \mathbb{Z}_2 \quad \text{with basis } \{z^1\}.$$

Note that $\{e^2 + e^3 + e^4 + e^5 + e^6\} = 2\{z^1\} = 0 \pmod 2$ (see the discussion before 8.2).

The definitions of *relative homology groups mod* 2 and *relative boundary homomorphisms mod* 2 are exactly parallel to those in Chapter 4, integer coefficients being replaced by coefficients in $\mathbb{Z}_2$. These relative groups are also vector spaces over $\mathbb{Z}_2$ and the homomorphisms are linear maps (see 8.2(2)). The statement and proof of 4.24 are still valid: given $c \in C_p(K,L;2)$ then $c \in Z_p(K,L;2)$ if and only if $\partial c \in C_{p-1}(L;2)$; and $c \in B_p(K,L;2)$ if and only if $\partial c' - c \in C_p(L;2)$ for some $c' \in C_{p+1}(K;2)$.

The homomorphisms i_*, j_* and ∂_* defined in 4.30, 4.32 and 4.34 carry over to the mod 2 situation and the *mod 2 homology sequence of a pair* (K, L) is exact. The statement and proof in 6.1 apply to the present case merely by changing integers to integers mod 2 throughout. The *reduced* sequence is also exact provided $L \neq \emptyset$.

The *excision theorem* is also true in mod 2 homology and the proof given in 6.4 carries over. The same is true of *collapsing* (6.6): if $K \searrow L$ then $i_* : H_p(L;2) \to H_p(K;2)$ is an isomorphism for all p. It follows from the proof of 6.8 that the mod 2 homology groups of any cone CK are $H_p(CK;2) = 0$, $p > 0$; $H_0(CK;2) \cong \mathbb{Z}_2$.

All the invariance properties of Chapter 5 carry over to the mod 2 case.

The calculation of $H_1(M;2)$ for a closed surface M is simpler than that of $H_1(M)$ since the analogue of 6.12 is:

Let K be obtained from M by removing a single 2-simplex. Then $i_* : H_1(K;2) \to H_1(M;2)$ is an isomorphism.

The proof is provided by the proof of 6.12(1), which carries over whether M is orientable or not. Hence we have the following result (compare 6.13).

Theorem 8.5. *Let M be a closed surface. Then $H_1(M;2)$ is as follows:*
(1) $H_1(M;2) = 0$ when M has standard form aa^1.
(2) $H_1(M;2) \cong (\mathbb{Z}_2)^k$ when M has standard form $a_1 a_1 \ldots a_k a_k$. A basis is $\{a_1\}, \ldots, \{a_k\}$.
(3) $H_1(M;2) \cong (\mathbb{Z}_2)^{2h}$ when M has standard form $a_1 b_1 a_1^{-1} b_1^{-1} \ldots a_h b_h a_h^{-1} b_h^{-1}$. A basis of $\{a_1\}, \{b_1\}, \ldots, \{a_h\}, \{b_h\}$. *(Cycles representing this basis are obtained by removing the arrows from those drawn in 6.14(2).)* □

Notice that $H_1(2nP;2) \cong H_1(nT;2)$ for any integer $n \geqslant 1$. Thus $2nP$ and nT have the same homology groups, mod 2, for any $n \geqslant 1$ (see 8.4(1)). Not surprisingly, the mod 2 groups are a weaker tool than the groups with integer coefficients.

Homology modulo 2   177

The theorem corresponding to the Euler characteristic formula of 6.16 is the following. Let K be a simplicial complex of dimension n. Recall that $\widehat{\beta}_p = \dim H_p(K; 2)$ and that $\alpha_p = $ number of p-simplexes of $K = \dim C_p(K; 2)$.

Theorem 8.6.

$$\chi(K) = \sum_{p=0}^{n}(-1)^p \alpha_p = \sum_{p=0}^{n}(-1)^p \widehat{\beta}_p.$$

The proof is simpler than that of 6.16 since it depends only on the 'rank plus nullity' theorem of linear algebra rather than on the more difficult theorem A.31. (The rank plus nullity theorem states that if $f : V \to W$ is a linear map between vector spaces, and V is finite dimensional, then $\dim(\operatorname{Im} f) + \dim(\operatorname{Ker} f) = \dim V$.) □

8.7. Alternating sum of dimensions theorem. *Suppose that*

$$0 \to A_n \to A_{n-1} \to \cdots \to A_1 \to A_0 \to 0$$

is an exact sequence of vector spaces and linear maps. Then

$$\sum_{i=0}^{n}(-1)^i \dim A_i = 0. \quad (\textit{Compare A.32(3).}) \qquad \square$$

Examples 8.8. (1X) Let K and L be as in 4.29(3) (Möbius band and its rim). Then $H_1(K, L; 2) \cong \mathbb{Z}_2$ with generator $\{e^1\}$ and $H_2(K, L; 2) \cong \mathbb{Z}_2$ with generator $\{t^1 + t^2 + t^3 + t^4 + t^5\}$. These follow either from a direct calculation as in 4.29(3) or from the mod 2 homology sequence of (K, L) and 8.4(5) above (the mod 2 homology groups of K). Compare 6.3(2), which does the calculations with ordinary homology.

(2X) Suppose that L is a connected subcomplex of K. Show that $j(Z_1(K; 2)) = Z_1(K, L; 2)$. Is this ever true if L is not connected? (Compare 4.29(4).)

(3X) Repeat Example 6.5 (removal of a 2-simplex from a two-dimensional simplicial complex) with homology mod 2.

(4X) Let S^n denote the set of proper faces of an $(n+1)$-simplex ($n \geqslant 1$). Then $H_p(S^n; 2) = 0$ ($p \neq 0, n$); $H_0(S^n; 2) \cong H_n(S^n; 2) \cong \mathbb{Z}_2$. (Compare 6.11(4).)

The statement and proof (7.1, 7.2) of the exactness of the *Mayer–Vietoris sequence* of a triad $(K; L_1, L_2)$ go over readily to the mod 2 case. Of course there is no longer any need for the minus sign in the definition of ϕ. All the examples of 7.4 work equally well with mod 2 homology groups (where $\mathbb{Z}$ must be changed to $\mathbb{Z}_2$ in (4)), and the reader is invited to try them. Likewise the *homology sequence of a triple* (M, N, K) is exact in mod 2 theory. (See 7.7, 7.8.)

Examples 8.9. (1X) Let M_1 and M_2 be closed surfaces and let K_1 and K_2 be obtained from them by removing a 2-simplex from each. Use the Mayer–Vietoris sequence (mod 2) to show that

$$i_* : H_1(K_\alpha; 2) \to H_1(M_\alpha; 2)$$

is an isomorphism for $\alpha = 1, 2$. Hence use the sequence (mod 2) to show that

$$H_1(M_1 \# M_2; 2) \cong H_1(M_1; 2) \oplus H_1(M_2; 2).$$

Compare 7.6(2)–(3).

(2) In the calculation, given in 7.5, of the first homology groups of triangulations of torus and Klein bottle by sticking together two cylinders, the only difference between the two cases is a difference of sign, and this disappears when we pass to mod 2 homology groups. Thus, as noted in 8.5 above, these two closed surfaces have isomorphic first homology groups.

(3X) It might be supposed that there was no way of distinguishing between an orientable and a non-orientable closed surface using exclusively mod 2 homology. This is not so. Let M be a closed surface and let M' be the first barycentric subdivision of M. Clearly M' is also a closed surface equivalent to M and M' is orientable if and only if M is orientable. Let w_p ($p = 0, 1, 2$) be the element of $C_p(M'; 2)$ given by the sum of all the p-simplexes of M'. It is straightforward to verify that w_1 is a 1-cycle mod 2 – this just uses the fact that every vertex of M' is an end-point of an even number of edges. In fact
$$\{w_1\} = 0 \Leftrightarrow M' \text{ (and hence } M \text{) is orientable.}$$

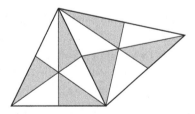

The reader may like to find an informal proof of this. The diagram is intended to be a hint.

Notice that w_0 and w_2 are also cycles. The elements $\{w_p\} \in H_p(M'; 2), p = 0, 1, 2$ are called the *Whitney* (or *Stiefel–Whitney*) *homology classes of* M. They are an example of 'characteristic classes' which are of great importance in algebraic topology. (See for example the books of Spanier and Milnor & Stasheff listed in the References. A modern reference for the homology classes mentioned above is Halperin & Toledo (1972). This article gives a proof of Whitney's original theorem, for which he did not publish a proof himself. There are many subsequent developments, for example Goldstein & Turner (1976) dispense with the need for taking a barycentric subdivision.)

★(4) If the ordinary homology groups $H_p(K)$ of a simplicial complex K are known, then it is actually possible to derive the groups $H_p(K;2)$. The formula is

$$H_p(K;2) \cong (H_p(K) \otimes \mathbb{Z}_2) \oplus (\text{Tor}(H_{p-1}(K), \mathbb{Z}_2)).$$

We shall not go into the details of the tensor product $\otimes$ and the torsion product Tor here, but the answer can be worked out for any given K using the following rules. (See for example the book of Hilton & Wylie listed in the References; note that they use ∗ for Tor. The formula is known as the *Universal Coefficients Theorem* for homology. See also the book *Elements of Homology Theory* by Prasolov.)

(i) Direct sums can be 'multiplied out', e.g.

$$(A_1 \oplus A_2) \otimes (B_1 \oplus B_2)$$
$$\cong (A_1 \otimes B_1) \oplus (A_1 \otimes B_2) \oplus (A_2 \otimes B_1) \oplus (A_2 \otimes B_2),$$

$$\text{Tor}(A_1 \oplus A_2, B_1 \oplus B_2)$$
$$\cong \text{Tor}(A_1, B_1) \oplus \text{Tor}(A_1, B_2) \oplus \text{Tor}(A_2, B_1) \oplus \text{Tor}(A_2, B_2);$$

(ii) $A \otimes B \cong B \otimes A$; $\text{Tor}(A, B) \cong \text{Tor}(B, A)$;
(iii) $\mathbb{Z} \otimes \mathbb{Z}_k \cong \mathbb{Z}_k$; $\mathbb{Z} \otimes \mathbb{Z} \cong \mathbb{Z}$; $A \otimes 0 = 0$;
  $\mathbb{Z}_m \otimes \mathbb{Z}_n \cong \mathbb{Z}_{(m,n)}$ where (m,n) = greatest common divisor of m and n;
(iv) $\text{Tor}(\mathbb{Z}, \mathbb{Z}_k) = 0$; $\text{Tor}(\mathbb{Z}, \mathbb{Z}) = 0$; $\text{Tor}(A, 0) = 0$; $\text{Tor}(\mathbb{Z}_m, \mathbb{Z}_n) \cong \mathbb{Z}_{(m,n)}$.

Thus for example take $K = P$, the projective plane, so that $H_0(P) \cong \mathbb{Z}$, $H_1(P) \cong \mathbb{Z}_2$, $H_2(P) = 0$. Then,

$$H_0(P;2) \cong (\mathbb{Z} \otimes \mathbb{Z}_2) \oplus \text{Tor}(0, \mathbb{Z}_2) \cong \mathbb{Z}_2$$
$$H_1(P;2) \cong (\mathbb{Z}_2 \otimes \mathbb{Z}_2) \oplus \text{Tor}(\mathbb{Z}, \mathbb{Z}_2) \cong \mathbb{Z}_2$$
$$H_2(P;2) \cong (0 \otimes \mathbb{Z}_2) \oplus \text{Tor}(\mathbb{Z}_2, \mathbb{Z}_2) \cong \mathbb{Z}_2.★$$

9
Graphs in surfaces

It has already been suggested, in the preamble to 4.19, that relative cycles have some connexion with the regions into which a graph divides a surface. To be a little more precise, let M be a closed surface and let K be a graph which is a subcomplex of M. If K is removed from M, what is left is a number of disjoint subsets of M which we refer to as the 'regions' into which K divides M. (Alternatively we may picture M as being cut along K: it falls into separate pieces, one for each region.) Consider one of these regions, and take out all the 2-simplexes from it – in the diagram one region is shaded and K drawn heavily. Now put back the 2-simplexes one by one. It is very plausible that the sum of the 2-simplexes replaced will have mod 2 boundary lying in K, i.e. will be an element of the relative cycle group $Z_2(M, K; 2)$, when all the 2-simplexes have been replaced and not before. Thus there may be some connexion between the number of regions and the dimension of $Z_2(M, K; 2) = H_2(M, K; 2)$.

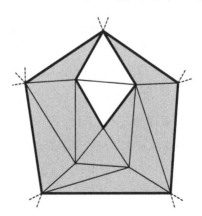

In order to begin establishing such a connexion we must have some interpretation of the number of regions in terms of homology theory. It is no good saying that since the regions are the 'components' of $M \setminus K$ the number of regions is the rank of '$H_0(M \setminus K)$' for $M \setminus K$ is not a simplicial complex unless $K = \emptyset$. We shall adopt a roundabout approach which replaces regions by subcomplexes of the second barycentric subdivision of M (see the discussion following 2.6).

Here is a rapid application of one of the results we are aiming for, namely that the number of regions is the dimension of $H_2(M, K; 2)$. Let M be a triangulation of the 2-sphere so that M has standard form aa^{-1} and $H_1(M; 2) = 0$, $H_2(M, 2) \cong \mathbb{Z}_2$ (see 8.4(1) and 8.5). Let K be a graph which is a subcomplex of M. Consider the homology sequence mod 2 of the pair (M, K):

$$H_2(K; 2) \to H_2(M; 2) \to H_2(M, K; 2) \to H_1(K; 2) \to H_1(M; 2)$$
$$\| \qquad \qquad \cong\| \qquad \qquad \qquad \qquad \qquad \qquad \|$$
$$0 \qquad \qquad \mathbb{Z}_2 \qquad \qquad \qquad \qquad \qquad \qquad 0$$

Now $H_1(K; 2)$ is isomorphic to the direct sum of the first homology groups of the components of K. If K_1 is such a component, the dimension of $H_1(K_1; 2)$ is $\mu(K_1) = \alpha_1(K_1) - \alpha_0(K_1) + 1$ (compare 8.4(2)). Hence the dimension of $H_1(K; 2)$ is $\alpha_1(K) - \alpha_0(K) + k$ where k is the number of components of K. Thus by the alternating sum of dimensions theorem (8.7)

$$r = \text{number of regions} = \alpha_1(K) - \alpha_0(K) + k + 1.$$

Thus

9.1 $\qquad \qquad \alpha_0(K) - \alpha_1(K) + r = k + 1.$

(Compare Euler's formula 1.27. Note that one of the ingredients of 1.27 was the Jordan Curve Theorem for polygons, so our result includes that theorem.)

If M were some closed surface other than the 2-sphere, $H_1(M; 2)$ would not be 0, and instead of 9.1 we should get

9.2 $\qquad \qquad \alpha_0(K) - \alpha_1(K) + r = k + 1 - d,$

where d is the dimension of the image of $i_* : H_1(K; 2) \to H_1(M; 2)$. The reader should check this informally for the three graphs in the torus drawn below.

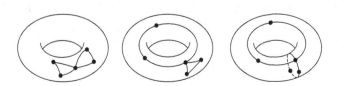

(In the right-hand diagram, for example, the two 1-cycles given by the loops on K are a basis for $H_1(M;2) \cong \mathbb{Z}_2 \oplus \mathbb{Z}_2$, so that $d = 2$ and 9.2 reads $5 - 6 + 1 = 1 + 1 - 2$.)

It is worth pointing out at this stage the connexion between 'graphs in the sphere' and 'graphs in the plane'. This must be to some extent an informal discussion since it uses Fáry's theorem (1.34) which was not proved in full detail. Let M be a sphere, that is a closed surface with standard form aa^{-1}, and let G be a graph which is a subcomplex of M. Possibly some barycentric subdivisions of M are necessary in passing to the standard form; remove one triangle from the subdivided M and proceed as in the diagrams. The result will be that some subdivision of G is realized as a curved graph in $\mathbb{R}^2$. Now use Fáry's theorem to realize the same subdivision of G as a straight graph in $\mathbb{R}^2$. Finally another application of the same theorem realizes G itself as a graph with straight edges in $\mathbb{R}^2$. Thus every graph in the sphere can be realized in the plane. A converse statement is also true – compare the diagrams in the discussion preceding 9.18, below.

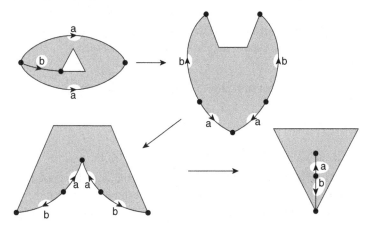

As was pointed out above, $M \setminus K$ is not usually a simplicial complex. We get around this difficulty by 'thickening' K a very little and then taking away not just K but the inside of the thickening as well.

Everything that follows can be generalized to the study of simplicial complexes K in 'combinatorial manifolds' M, which are higher-dimensional analogues of surfaces, though the results are not usually as complete in higher dimensions as they are here. One such generalization is given in 9.18 where we outline a proof that non-orientable closed surfaces cannot be realized in $\mathbb{R}^3$. Information on higher dimensions can be found in, for example, the books of Prasolov listed in the References.

Regular neighbourhoods

Let M be a closed surface and K a subcomplex of M; for the moment K need not be a graph. Let M'' and K'' be the second barycentric subdivisions of M and K respectively, so that K'' is a subcomplex of M''. Consider the following subcomplexes of M'':

9.3 $\begin{cases} N = \text{Closure } \{s \in M'' : s \text{ has at least one vertex in } K''\} \\ V = \{s \in M'' \ : \ s \text{ has no vertex in } K''\}. \end{cases}$

(Thus N is the closed star of K'' in M'' – see 3.24.) N is called a *regular neighbourhood* of K in M, and it surrounds the various components of K in such a way that the pieces of N surrounding different components do not intersect. (See the diagram on the next page.)

This non-intersection property would not be guaranteed by taking only first barycentric subdivisions, as the following example shows. In the diagram, $(v^0 v^1 v^2)$ is a triangle of M: suppose that K consists of $(v^0), (v^1), (v^2)$ and $(v^1 v^2)$. Then replacing M'' and K'' by M' and K' in the definition of N above, the pieces of N surrounding the two components of K intersect along ABC.

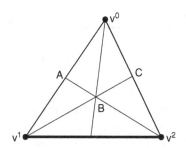

Now return to the genuine N defined by 9.3. Since N keeps as close as possible to K – no vertex of N is more than 'one step away' from K'' along the edges of M'' – the following are plausible:

(i) N and K have the same number of components (this follows from 9.5 below);

(ii) each region contains precisely one component of V (this component is a slightly shrunk down version of the region).

(Indeed taking a component V_i of V we can consider the subset of M'' defined by

$$V_i^+ = V_i \cup \{s \in M'' : s \text{ has a vertex in } K'' \text{ and a vertex in } V_i\},$$

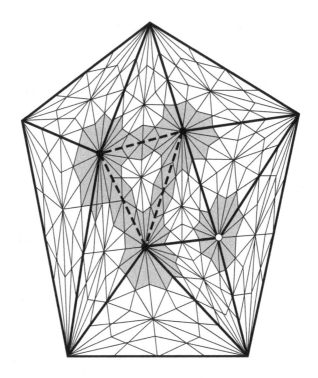

The diagram shows 11 triangles of a closed surface M, each subdivided into 6 triangles of M' and 36 triangles of M''. K is the subcomplex of M consisting of the 3 broken edges and their end-points, together with the isolated white vertex of M. Thus K is automatically subdivided into K' and K''. The shaded triangles and their faces make up the *regular neighbourhood* N of K in M: each shaded triangle has at least a vertex in common with K''. Note that N has two components: each surrounds a component of K. The white triangles and their faces make up the 'complementary' complex V: thus $N \cap V$ is 3 simple closed polygons.

and call this, or possible $|V_i^+|$, the *region determined by* V_i. It is not hard to prove that the V_i^+ are disjoint and that their union is $M \setminus K$. But they are not subcomplexes of M'' (unless $K = \emptyset$ or $K = M$) and that is the reason for using the V_i instead. As we shall see, the V_i have pleasant properties – they are 'surfaces' – which makes them comparatively easy to study.)

In view of (ii) and the fact that V is a simplicial complex we now have a measure of the number of regions, and we frame this as a definition.

Definition 9.4. Let V be as in 9.3. The *number of regions* into which K divides M is the number of components of V.

Regular neighbourhoods 185

The object, then, is to calculate $H_0(V;2)$ (or $H_0(V)$); for good measure we shall also calculate $H_1(V;2)$, and, when M is orientable, $H_1(V)$. This gives additional information about the regions, as we shall see. The first step is the following result.

Lemma 9.5. *N collapses to K''. Hence $i_* : H_p(K'') \to H_p(N)$ is an isomorphism for each p, and $H_p(N, K'') = 0$ for each p. The corresponding results with mod 2 homology groups also hold.*

Proof Consider a 2-simplex $(v^0 v^1 v^2)$ of the first barycentric subdivision M' of M. The number of vertices of this 2-simplex which belong to K' is either 0, 1, 2 or 3, and the corresponding intersections of N with $(v^0 v^1 v^2)$ are shaded in the top line of the diagrams below. (Compare the diagram on the previous page.) The lower diagrams show, in the second and third cases, the result of collapsing the 2-simplexes and a 1-simplex of $N \backslash K''$ which lie inside $(v^0 v^1 v^2)$. In this way N collapses to K'' together with some 1-simplexes of N, and a similar consideration of the 1-simplexes of M' shows that these 1-simplexes of $N \backslash K''$ can also be collapsed away. Hence $N \searrow K''$

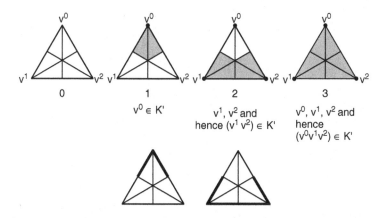

The remaining statements of the lemma follow from the invariance of homology groups under collapsing (6.6), the homology sequence of (N, K'') (compare 6.3(5)) and the corresponding the results for homology groups mod 2. □

Corollary 9.6. $H_p(M'', K'') \cong H_p(M'', N)$ *for all p. The same result holds using homology mod 2.*

Proof This is an immediate consequence of the exactness of the homology sequence of the triple (M'', N, K'') and 9.5. (See 7.8, and replace M by M'', K by K''.) Alternatively a direct argument can be given, and we sketch one here. As usual, all pth chain groups are assumed to be subgroups of $C_p(M'')$.

Let $\beta : H_p(M'', K'') \to H_p(M'', N)$ be defined by $\beta\{z\} = \{jz\}$, where $z \in Z_p(M'', K'')$ and $j : C_p(M'') \to C_p(M'', N)$ is the usual homomorphism which 'forgets N'. It is straightforward to show that β is well-defined and a homomorphism; we need to show that it is an isomorphism. This will use $H_{p-1}(N, K'') = 0 = H_p(N, K'')$.

To show β is monic (injective), suppose $\{jz\} = 0$, i.e $jz = \partial x + y$ where $x \in C_{p+1}(M'')$ and $y \in C_p(N)$. (Compare 4.24(2).) We can clearly assume $x \in C_{p+1}(M'', N)$ by adjusting y; let us also write $y = y_1 + y_2$ where $y_1 \in C_p(N, K'')$ and $y_2 \in C_p(K'')$. In the diagram, $y_2 = 0$. Further write $\rho z \in C_p(N, K'')$ for the part of z lying in N; thus $z = jz + \rho z$.

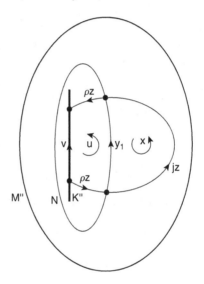

Now

$$\rho z + y_1 = z - \partial x - y_2 \in C_p(N, K'') \text{ and } \partial(\rho z + y_1) = \partial z - 0 - \partial y_2 \in C_{p-1}(K'')$$

so that

$$\rho z + y_1 \in Z_p(N, K'') = B_p(N, K'') \text{ since } H_p(N, K'') = 0.$$

Thus $\rho z + y_1 = \partial u + v$ say where $u \in C_{p+1}(N, K'')$, $v \in C_p(K'')$. As the diagram suggests, to prove $\{z\} = 0$ we use $u + x$; certainly $u + x \in C_{p+1}(M'', K'')$, and

$$\partial(u + x) = \rho z + y_1 - v + jz - y_1 - y_2 = z - (v + y_2).$$

Since $v + y_2 \in C_p(K'')$ this shows that $z \in B_p(M'', K'')$, as required.

The proof that β is epic (surjective) is easier, and depends on taking $z \in Z_p(M'',N)$ and writing $\partial z = x+y$ where $x \in C_{p-1}(N,K'')$ and $y \in C_{p-1}(K'')$. Then $x \in Z_{p-1}(N,K'') = B_{p-1}(N,K'')$ and writing $x = \partial u + v$ in an analogous way to that above, it turns out that $\beta\{z-u\} = \{z\}$. The reader should try drawing a diagram of this in the case $p=1$. □

Now $M'' = N \cup V$, so that by excision (6.4) $H_p(M'',N) \cong H_p(V, N \cap V)$ for all p. Also $H_p(M'', K'') \cong H_p(M, K)$ by the invariance of homology groups under barycentric subdivision (5.7, 5.8), so that we can combine the foregoing results into the following proposition.

Proposition 9.7. $H_p(M,K) \cong H_p(V, N \cap V)$ for all p; $H_p(M,K;2) \cong H_p(V, N \cap V; 2)$ for all p. □

In order to identify the groups $H_p(V, N \cap V)$ we must digress a little on the subject of surfaces.

Surfaces

The class of surfaces includes that of closed surfaces (Chapter 2) and also includes such objects as triangulations of a Möbius band which fail to be closed surfaces because there exist vertices (on the rim) whose link is not a simple closed polygon but a simple polygonal arc. (See 2.2(6).)

Definition 9.8. A *surface* is a connected simplicial complex of dimension 2 in which the link of each vertex is either a simple closed polygon or a simple polygonal arc.

It is not difficult to see from this that every edge of a surface M is a face of either one or two triangles of M (compare 2.2(2)). For example if an edge were a face of three triangles then the link of either end-point of the edge would contain ⋎ which cannot be part of any simple closed polygon or simple polygonal arc.

Definition 9.9. A *boundary* ∂M of a surface M is the closure of the set of edges of M which are a face of exactly one triangle of M. (Thus ∂M consists of these edges of M and their end-points, and is a subcomplex of M.) Note that a vertex v of M belongs to ∂M if and only if the link of v in M is a simple polygonal arc.

A surface is called *orientable* if its triangles can be oriented in such a way that two triangles with a common edge are always oriented coherently. (Compare 2.3.) Otherwise it is *non-orientable*.

Proposition 9.10. (1) *For any surface M, the boundary ∂M consists of a collection (possibly empty) of simple closed polygons.*

(2) *Given any two triangles t^1, t^2 of M there is a chain of triangles connecting them, any two consecutive members of the chain having an edge in common.*

Proof

(1) We need only verify that any vertex v of ∂M is an end-point of precisely two edges of ∂M. The link of v in M is a simple polygonal arc, with end-vertices u and u' say, so that (vu) and (vu') both belong to ∂M. If (vw) is a further edge of M with end-point v then by definition $w \in \text{Lk}(v, M)$; this implies that $(vw) \notin \partial M$ since (vw) will be a face of two triangles of M.

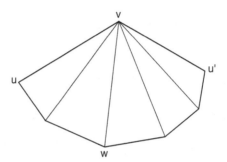

(2) The argument suggested in 2.2(5) for the case of closed surfaces still works, since it only uses the fact that the link of each vertex is connected. □

Examples 9.11. (1) Any closed surface is a surface with empty boundary.

(2) The triangulations of Möbius band and cylinder in 3.22(3) are surfaces. The boundaries consist of one and two simple closed polygons respectively.

(3X) Let M be obtained from a closed surface by removing a single triangle. Then M is a surface. What happens if two triangles are removed?

(4X) Let M be a surface with boundary ∂M consisting of $n \geqslant 1$ simple closed polygons. By adding a cone on each boundary component, the cones having different vertices and intersecting M precisely along the components of ∂M, we obtain a closed surface $\overline{M}$. The link of a vertex of ∂M is 'completed' to a simple closed polygon by the addition of two edges and a vertex of a cone, while the link of the vertex of one of the cones is a component of ∂M, which is a simple closed polygon. (Compare 6.11(1), (2).) Thus a surface can be thought of as a 'closed surface with disks cut out'.

The *genus* of M is defined to be the genus (number of handles or crosscaps, see 2.8) of $\overline{M}$. Thus the genus of a cylinder is 0 and the genus of a Möbius band is 1. Purists may like to use this construction to define cylinders and Möbius bands: a surface M is a (triangulation of a) cylinder if it has two boundary components and $\overline{M}$ has standard form aa^{-1}, etc.

Show that $\chi(\overline{M}) = \chi(M) + n$ (see the second formula (Euler characteristic of a union) of 6.17(1) and recall that a cone collapses to a point (6.8) and therefore has Euler characteristic 1). Show that M is orientable if and only if $\overline{M}$ is orientable.

(5) According to 3.35(3), a subcomplex of a closed surface which does not contain all the triangles of the closed surface collapses to a subcomplex of the 1-skeleton. Applying this to the subcomplex M of the closed surface $\overline{M}$ constructed in (4) this shows that M collapses to a graph. Hence: for any surface M with non-empty boundary, $H_2(M) = 0$ and $H_1(M)$ is free abelian. A graph to which M collapses is sometimes called a *spine* of M.

(6X) Let M be a surface as in (4) and let M_1 and M_2 be surfaces each isomorphic (as simplicial complexes) to M. Form the simplicial complex DM (a 'double' of M) from the disjoint union $M_1 \cup M_2$ by identifying each oriented simplex of ∂M_1 with the corresponding oriented simplex of ∂M_2. Thus starting with a cylinder the picture is as follows:

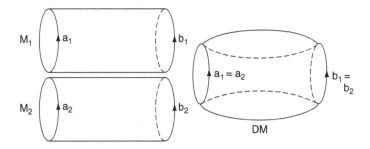

In this case DM is a torus. Show that DM is always a closed surface, orientable if and only if M is orientable.

Show that $\chi(DM) = 2\chi(M)$. As in (4), we take the genus of M to be that of $\overline{M}$; find the genus of DM in terms of that of M. (Consider separately the two cases M orientable and M non-orientable.)

It is also possible to formulate a definition of DM in which oriented boundary components are identified with opposite orientations (in fact this is the more usual definition). This requires a more complicated definition, but granted that it can be made does the change affect the genus of DM?

It was proved in (5) above that $H_2(M) = 0$ for any surface with non-empty boundary. It follows similarly that $H_2(M; 2) = 0$ in the same circumstances.

A more interesting group to consider is $H_2(M, \partial M)$, for when M is orientable the sum of all 2-simplexes, coherently oriented, is a non-zero element of this group. In fact the proof (4.18) in the case of a closed surface can be imitated here to show $H_2(M, \partial M) \cong \mathbb{Z}$ with the element already mentioned as generator. (A more precise definition of this generator is given by the formula in 4.18.) In the non-orientable case $H_2(M, \partial M) = 0$ (as for closed surfaces); on the other hand with homology mod 2 the sum of all triangles always generates $H_2(M, \partial M; 2) \cong \mathbb{Z}_2$. These results are collected in the following proposition.

Proposition 9.12. *Let M be a surface with non-empty boundary ∂M.*

	$H_2(M)$	$H_2(M;2)$	$H_2(M,\partial M)$	$H_2(M,\partial M;2)$
M orientable	0	0	$\mathbb{Z}$	$\mathbb{Z}_2$
M non-orientable	0	0	0	$\mathbb{Z}_2$

(See 4.18 for the case $\partial M = \emptyset$.) □

Proposition 9.13. (1) *Let M be a surface. Then*

$$H_1(M; 2) \cong H_1(M, \partial M; 2).$$

(2) *Let M be an orientable surface. Then*

$$H_1(M) \cong H_1(M, \partial M).$$

Proof (1) Consider the homology sequence, mod 2, of $(M, \partial M)$, and assume $\partial M \neq \emptyset$ since otherwise there is nothing to prove. (The notation ';2' is omitted here to save space.)

$$H_2(M) \to H_2(M, \partial M) \to H_1(\partial M) \to H_1(M) \to$$
$$ 0 1 n$$
$$H_1(M, \partial M) \to H_0(\partial M) \to H_0(M) \to H_0(M, \partial M)$$
$$ n 1 0$$

The numbers are the dimensions of the vector spaces in the sequence, n being the number of components of ∂M ($n \geqslant 1$). By the alternating sum of dimensions theorem (8.7), $H_1(M; 2)$ and $H_1(M, \partial M; 2)$ have the same dimension, and so are isomorphic.

(2) A similar argument using the homology sequence with $\mathbb{Z}$ coefficients and the alternating sum of ranks theorem (A.32(3)) shows that $H_1(M)$ and

$H_1(M, \partial M)$ have equal rank. Now $H_1(M)$ is free abelian, by 9.11(5), so the result will follow once it is known that $H_1(M, \partial M)$ is free abelian. Let $\overline{M}$ be constructed as in 9.11(4) and let C be the union of the n cones on boundary components of M. By excision, $H_1(\overline{M}, C) \cong H_1(M, \partial M)$. Now from the reduced homology sequence of the pair $(\overline{M}, C)$ and the fact that $H_1(C) = 0$ it follows that $H_1(\overline{M}, C) \cong H_1(\overline{M}) \oplus \mathbb{Z}^{n-1}$. The result now follows by using 6.13, which say that $H_1(\overline{M})$ is free abelian. □

Remarks 9.14. (1) The proof of 9.13 is, to say the least, unnatural. A more illuminating version of 9.13 can be given using 'cohomology' groups, which we do not use in this book – though 'cochains' are mentioned in the appendix to Chapter 1. The version given here will suffice for our purposes.

(2X) The homomorphism j_* between the first and second groups in each part of 9.13 is not in general an isomorphism. In fact using information already given, and the homology sequence of $(M, \partial M)$, the reader should be able to prove the following. Assume $\partial M \neq \emptyset$, then:

$j_* : H_1(M; 2) \to H_1(M, \partial M; 2)$ is an isomorphism if and

only if ∂M is connected ($n = 1$).

$j_* : H_1(M) \to H_1(M, \partial M)$ is an isomorphism if and only if

M is orientable and ∂M is connected.

Lefschetz duality

Now we return to the relative homology groups of $(V, N \cap V)$ which appear in 9.7, and justify the digression on surfaces.

Proposition 9.15. *In the notation of 9.3, each component of V is a surface whose boundary is the intersection of the component with N. Provided $K \neq \emptyset$ and $K \neq M$ this boundary is never empty. The same result holds with V and N interchanged.*

Proof We shall prove the first two sentences and leave the third as an (easy) exercise. (The reader may find the diagram on page 184 helpful in following the proof.) For the first sentence we need to show the following: if v is a vertex of V then

(i) $v \notin N \Rightarrow \mathrm{Lk}(v, V)$ is a simple closed polygon

(ii) $v \in N \Rightarrow \mathrm{Lk}(v, V)$ is a simple polygonal arc.

(i) Suppose $v \notin N$. Certainly $\mathrm{Lk}(v, M'')$ is a simple closed polygon, for M'' is a closed surface. Furthermore $\mathrm{Lk}(v, V) \subset \mathrm{Lk}(v, M'')$ so we just need to prove the converse inclusion. For this suppose that $s \in \mathrm{Lk}(v, M'')$, so that $sv \in M''$. If a vertex of s failed to belong to V then it would belong to K'' and so sv and hence v would belong to N, contrary to assumption. Thus s, and so sv, belong to V, and this shows $s \in \mathrm{Lk}(v, V)$.

(ii) Suppose $v \in V \cap N$. Then $v \notin K''$ but there is a 1-simplex $(vw) \in M''$ with $w \in K''$.

Suppose v is a vertex of M'. Then w, being one step away from v along an edge of M'', is the barycentre of a simplex $s \in M'$ of dimension 1 or 2. Also $s \in K'$ since $w \in K''$ (recall w is a vertex of K'' if and only if w is the barycentre of a simplex of K'). Finally v is a vertex of s so $v \in K'$, a contradiction.

Hence v must be the barycentre of a 1- or 2-simplex of M' and we take these cases in turn.

Case 1 $v = \hat{s}$ (barycentre of s) where s is a 1-simplex of M'. From the left-hand diagram below, $\mathrm{Lk}(v, M'')$ has four vertices $\hat{a}, \hat{b}, \hat{c}, \hat{d}$. Now if $a \in K'$ then $s \in K'$ so $v \in K''$, a contradiction. Hence $a \notin K'$, i.e. $\hat{a} \notin K''$, i.e. $\hat{a} \in V$. Similarly $\hat{c} \in V$. Thus the w mentioned above must be either $\hat{b}$ or $\hat{d}$, say $\hat{b}$ which then belongs to K'. If $\hat{d}$ also belonged to K' then $s = (\hat{b}\hat{d})$ would belong to K' (the simplexes of K' are precisely those of M' all of whose vertices belong to K') and so $v \in K''$, a falsehood. The conclusion is that $\mathrm{Lk}(v, V)$ is as in the right-hand diagram below.

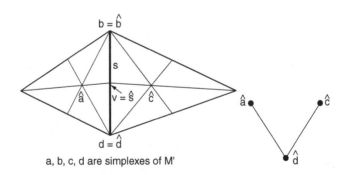

a, b, c, d are simplexes of M'

Case 2 $v = \hat{s}$ where s is a 2-simplex of M'. Let $s = (v^1 v^2 v^3)$, as in the left-hand diagram below. Now at most two of v^1, v^2, v^3 can belong to K' since if all three did then s would belong to K' (same reason as in Case 1), and so v would belong to K''. If, say, v^1 and v^2 belong to K', then just v^3, v^4, v^5 belong to V and $\mathrm{Lk}(v, V)$ is as in the middle diagram; if on the other hand just $v^1 \in K'$ then $\mathrm{Lk}(v, V)$ is as in the right-hand diagram.

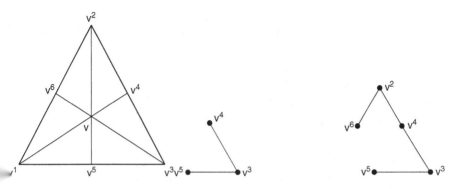

Finally we must check that, for any component V_1 of V, $V_1 \cap N$ is non-empty, assuming $K \neq \emptyset$ and $K \neq M$. Now these two restrictions on K imply, respectively, that $N \neq \emptyset$ and $V \neq \emptyset$, so choose a vertex $v \in V_1$ and a vertex $w \in N$ and join them by a path $v \ldots w$ in M''. Let w' be the first vertex in this path which belongs to N; then all previous vertices, and hence edges, in the path belong to V, and therefore to V_1. Furthermore $w' \in V$ since otherwise $w' \in K''$ and this implies that the vertex preceding w' in the path belongs to N. Hence $V_1 \cap N \neq \emptyset$. □

Remarks 9.16. (1) The above analysis applies to a subcomplex K of a closed surface M. It is not difficult to generalize it to the case where M has non-empty boundary. The definitions of V and N remain the same, and 9.15 remains true except that the boundary of a component of V is the intersection of the component with $(N \cup \partial M)$. The situation is indicated in the diagram below. Likewise each component of N is a surface with boundary the intersection of the component with $(V \cup \partial M)$.

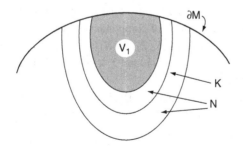

(2) In the notation introduced before 9.4, let V_1^+ be the region determined by the component V_1 of V. Then in fact $|V_1^+|$ is homeomorphic to $|V_1 - \partial V_1|$: the region can be continuously shrunk down to the interior of the surface V_1. In particular if V_1 is a triangulation of a disk (i.e. of the plane region bounded

by a simple closed polygon) then $|V_1^+|$ is homeomorphic to the open disk $\{(x, y) \in \mathbb{R}^2 : x^2 + y^2 < 1\}$. We say then that 'the region is an open disk'. In fact this happens if and only if $H_1(V_1; 2) = 0$ (see 9.21). Compare the discussion of triangulating a disk in Chapter 5, preceding 5.10.

Of course the closure $\text{Cl}(V_1^+)$ is a subcomplex of M''. Note that it is not necessarily a surface and $|\text{Cl}(V_1^+)|$ is not necessarily homeomorphic to $|V_1|$. An example appears in the first diagram of this chapter.

Theorem 9.17. *(A version of 'Lefschetz duality'.)*

(1) *Let M be a closed surface and K a subcomplex of M, and let V be given by 9.3. Then*

$$H_0(V; 2) \cong H_2(M, K; 2) \quad \text{and} \quad H_1(V; 2) \cong H_1(M, K; 2).$$

In particular the number of regions into which K divides M is $\dim(H_2(M, K; 2))$.

(2) *Let M be an orientable closed surface and K a subcomplex of M. Then*

$$H_0(V) \cong H_2(M, K) \quad \text{and} \quad H_1(V) \cong H_1(M, K).$$

In particular the number of regions into which K divides M is $\text{rank}(H_2(M, K))$.

Proof (1) By 9.7, $H_2(M, K; 2) \cong H_2(V, N \cap V; 2)$. But V is a disjoint union of surfaces and $N \cap V$ is the union of their boundaries, by 9.15. Thus by 9.12 (and 7.4(1) which gives the homology of a disjoint union), $H_2(V, N \cap V; 2)$ has dimension equal to the number of components of V, and hence is isomorphic to $H_0(V; 2)$. Likewise, $H_1(M, K; 2) \cong H_1(V, N \cap V; 2)$ by 9.7 and this is isomorphic to $H_1(V; 2)$ by 9.13(1) (and 7.4(1)) since V is a union of surfaces as above.

(2) The proof is parallel to that of (1), using the fact that since M is orientable so is M'' and all components of V (and N). □

Using 'cohomology' the Lefschetz duality theorem, like 9.13, can be given a more natural look. In particular it is possible to describe explicit isomorphisms between the groups involved. Several books give the details (which require much more machinery than has been set up here). See for example the textbooks of Spanier or Prasolov (*Elements of Homology Theory*) listed in the References.

★ A three-dimensional situation

Before going on to applications of the Lefschetz duality theorem proved above here is an informal application of the corresponding theorem in the next dimension up. We shall not prove this theorem – though a proof closely analogous to that for 9.17 can be constructed.

The *3-sphere* is the set of points in $\mathbb{R}^4$ with coordinates (x_1, x_2, x_3, x_4) satisfying $x_1^2 + x_2^2 + x_3^2 + x_4^2 = 1$. A particular triangulation of the 3-sphere is the simplicial complex S^3 considered in 6.11(4): the set of proper faces of a 4-simplex. Let us assume that any triangulation M of the 3-sphere will have homology groups isomorphic to those of S^3 (this is a consequence of the invariance theorem 5.13), i.e. $H_0(M) \cong \mathbb{Z}, H_1(M) = 0 = H_2(M), H_3(M) \cong \mathbb{Z}$ and the same with $\mathbb{Z}_2$ coefficients except that the $\mathbb{Z}$'s are replaced by $\mathbb{Z}_2$'s.

Now let K be a subcomplex of M. Then it can be shown that the number of (three-dimensional) regions into which K divides M is the dimension of $H_3(M, K; 2)$. It is also equal to the rank of $H_3(M, K)$, since M is orientable in a sense analogous to that for surfaces.

I claim the following: if a simplicial complex K can be realized in $\mathbb{R}^3$ then it is (isomorphic to) a proper subcomplex of some triangulation M of the 3-sphere. Here is an informal proof that this is so. Suppose K is realized in $\mathbb{R}^3$; then enclose it inside a big tetrahedron and triangulate the (solid) tetrahedron with K as a subcomplex (this can always be done). Now regard the solid tetrahedron as one three-dimensional face of a 4-simplex and triangulate the 3-skeleton S^3 of the 4-simplex so that the triangulated solid tetrahedron is a subcomplex. This makes K a subcomplex of a subdivision of S^3. The sequence of steps one dimension down is illustrated below.

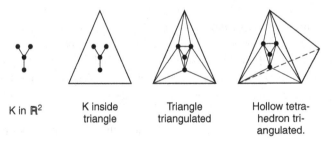

K in $\mathbb{R}^2$ | K inside triangle | Triangle triangulated | Hollow tetrahedron triangulated.

(In this example there is no need to divide up the other three faces of the hollow tetrahedron.)

Suppose now that K is a non-orientable closed surface and is a subcomplex of M. Consider the exact sequence

$$H_3(K) \to H_3(M) \to H_3(M, K) \to H_2(K) \to H_2(M).$$

Since $H_2(K) = 0$ and $H_3(K) = 0$ this gives the rank of $H_3(M, K)$ as one: there is one region, i.e. K does not 'separate' M into two or more regions. The same sequence with $\mathbb{Z}_2$ coefficients, using $H_2(K; 2) \cong \mathbb{Z}_2$, $H_3(K; 2) = 0$ gives the dimension of $H_3(M, K; 2)$ as two: there are two regions. This contradiction shows that K cannot be a subcomplex of M. Using the result above, this gives:

Theorem 9.18. *Let K be a non-orientable closed surface. Then K cannot be realized in $\mathbb{R}^3$.* □

We have seen, in 2.17(5), that for every orientable closed surface there is an equivalent closed surface realizable in $\mathbb{R}^3$. It follows from the exact sequence used above that, when an orientable closed surface is realized in $\mathbb{R}^3$, it separates $\mathbb{R}^3$ into two regions (the regions 'inside' and 'outside' the surface).

An orientable closed surface K in $\mathbb{R}^3$ is also said to be two-sided. To explain the meaning of this term take a point x, interior to one of the triangles of K, and construct a very short segment $x'x''$ normal (i.e. perpendicular) to the triangle and with x as its mid-point. Thus x' and x'' are 'on opposite sides' of K. We call K *two-sided* in $\mathbb{R}^3$ if it is impossible to go from x' to x'' along a path which always keeps very close to K and never crosses K. Clearly if such a path existed then K could not separate $\mathbb{R}^3$ into two regions, so indeed an orientable closed surface in $\mathbb{R}^3$ is two-sided. (The definition can be made precise in terms of regular neighbourhoods; we really want K as a subcomplex of a triangulation of the 3-sphere, and then require that the boundary of a regular neighbourhood is not connected.)

Some people say that a Klein bottle in $\mathbb{R}^3$ is *one-sided* (i.e. not two-sided), but as a Klein bottle cannot be realized in $\mathbb{R}^3$ this statement must be treated with caution. Perhaps the reader can decide for himself whether he wants to call the 'surface' with self-intersection drawn in 2.2(4) one-sided or not. One-sidedness of a surface in $\mathbb{R}^4$ is not possible, and a surface cannot separate $\mathbb{R}^4$.

It does make good sense to say that a Möbius band in $\mathbb{R}^3$ is one-sided and a cylinder in $\mathbb{R}^3$ is two-sided, but the interpretation of these statements needs a little care on account of the boundary rims. The path from x' to x'' has to keep away from the rims.

It is possible to realize closed surfaces inside other spaces besides euclidean spaces. Consider first a solid ball (a 2-sphere plus all the points inside it in $\mathbb{R}^3$) and form a new object by identifying together each pair b, b' of diametrically opposite points on the boundary sphere. The result is called 'three-dimensional projective space' by analogy with the projective plane (see 2.7(3)). Indeed the equatorial disk (shaded in the diagram below) turns into a projective plane since opposite points on its boundary circle are identified. Thus the projective plane is realized inside three-dimensional projective space, and it is easy to see that it is one-sided (and hence cannot separate the space).

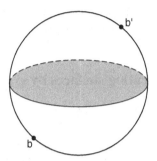

The three-dimensional projective space is an example of a *three-dimensional manifold* in which the points close to any given point p form a 'neighbourhood' of p which is essentially a solid ball – compare the corresponding idea for closed surfaces, where 'solid ball' is replaced by 'disk', in the discussion at the beginning of Chapter 2. For points $b = b'$ as above the neighbourhood of $b = b'$ is in two parts before identifications take place, but these fit together to form a solid ball when diametrically opposite points are all identified in pairs.

Another example of a three-dimensional manifold is obtained by starting with a solid torus (a torus T plus all the points inside it in $\mathbb{R}^3$) and identifying together each pair c, c' of points which are diametrically opposite on any of the 'vertical' circles lying on T. Call the result M. Thus each vertical disk, such as the one shaded, turns into a projective plane P in M. It is not difficult to see that P (which is non-orientable) is two-sided in M; however P does not separate M. The torus T also turns into a closed surface T' in M. If we slice T into two equal pieces by a horizontal plane then T' is obtained from one of theses pieces by identifying together the two end-points on each vertical semicircle. Thus T' is another torus, and in fact the orientable surface T' is one-sided in M (and hence cannot separate M into two regions).

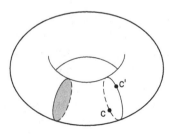

198   Graphs in surfaces

In fact, by taking suitable three-dimensional manifolds, any combination of orientable/non-orientable, two-sided/one-sided, separates/does not separate the manifold, is possible, except that, as pointed out above, a surface which separates is always two-sided. (That leaves six combinations; can the reader find a non-orientable surface which does separate, and so is two-sided?) ★

Separating surfaces by graphs

Let K be a graph which is a subcomplex of a closed surface M. We can ask the question: when will K *separate* M, i.e. when will it divide M into r regions where $r > 1$? This can be thought of more picturesquely as follows: Suppose we cut M along the edges of K; when will M fall apart? Imagining that our scissors are rather blunt we can even picture them removing a regular neighbourhood of K so that M drops into a collection of surfaces $V_1, \ldots, V_r$, the components of V.

First let K be a simple closed polygon, so that $H_1(K; 2) \cong \mathbb{Z}_2$. Consider the exact sequence

$$H_2(K;2) \to H_2(M;2) \to H_2(M,K;2) \to H_1(K;2) \xrightarrow{i_*} H_1(M;2).$$
$$\quad\; 0 \qquad\qquad\;\; 1 \qquad\qquad\;\;\; r \qquad\qquad\;\;\; 1$$

The numbers are dimensions of homology groups, using 9.17. If $i_* = 0$ then exactness implies that $r = 2$; if $i_* \neq 0$, so that the 1-cycle mod 2 on M given by the simple closed polygon K is not homologous to zero, then exactness implies that $r = 1$. Hence:

Proposition 9.19. *A simple closed polygon separates a closed surface if and only if the 1-cycle mod 2 given by the polygon is homologous to zero on the closed surface. (Compare 8.4(3).)*

A general graph K will separate M if and only if K contains a simple closed polygon of the same kind. □

Now let us ask what is the largest graph K which can be embedded in a closed surface M without separating it. Here we allow M to be replaced by any equivalent closed surface (one with the same standard form), so that we are asking questions like: of all the graphs which are subcomplexes of some triangulation of a torus and which fail to separate, which are the 'largest'? (Compare the discussion following 2.20.) As a measure of largeness we take $\dim H_1(K; 2)$, which is the sum of the cyclomatic numbers of the components of K. Consider the exact sequence, where $\mathbb{Z}_2$ coefficients are assumed:

$$H_2(K) \to H_2(M) \xrightarrow{j_*} H_2(M,K) \xrightarrow{\partial_*} H_1(K)$$
$$\xrightarrow{i_*} H_1(M) \to H_1(M,K) \to \widetilde{H}_0(K).$$

Now $H_2(K; 2) = 0$, $H_2(M; 2) \cong \mathbb{Z}_2$.
Suppose that K does not separate M. By Lefschetz duality (9.17) $H_2(M,K; 2) \cong \mathbb{Z}_2$ and j_*, being a monomorphism (injective), must be an isomorphism, so that $\partial_* = 0$ and i_* is monic (also injective). Hence

$$\dim H_1(K; 2) \leqslant \dim H_1(M; 2), \text{ with equality if and only}$$

if i_* is an isomorphism.

Conversely if i_* is an isomorphism then $\partial_* = 0$, so that j_* is an isomorphism and $H_2(M,K; 2) \cong \mathbb{Z}_2$. Hence:

Proposition 9.20. *K is a maximal non-separating graph for M (maximal in the sense of having the largest possible* $\dim H_1(K; 2)$*) if and only if* $i_* : H_1(K; 2) \to H_1(M; 2)$ *is an isomorphism.* □

Examples of maximal non-separating graphs on a torus and a double-torus are indicated below. (Cycles mod 2 representing a basis for $H_1(M; 2)$, when M is orientable, are obtained by removing the arrows from the diagram in 6.14(2).)

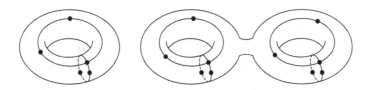

If we suppose that K is connected, then $\widetilde{H}_0(K; 2) = 0$ and i_* is an isomorphism implies that $H_1(M,K; 2) = 0$. Since in this case V is connected, Lefschetz duality implies that $H_1(V; 2) = 0$. What does this tell us about V? It says, in the first place, that ∂V is connected, for using the mod 2 homology sequence of $(V, \partial V)$ we find $H_1(\partial V; 2) \cong \mathbb{Z}_2$. The corresponding closed surface $\overline{V}$ (9.11(4)) is therefore the union of V with a single cone, so that $\chi(\overline{V}) = \chi(V) + 1 = 2$ since $\chi(V)$ is the alternating sum of the dimensions of the mod 2 homology groups of V (8.6) and this is 1. Now the only closed surfaces with $\chi = 2$ are triangulations of a 2-sphere (standard form aa^{-1}), so $\overline{V}$ is one of these. Returning to V amounts to removing the star of one vertex from a closed surface with standard form aa^{-1}, and this gives a triangulation of a disk (the plane region bounded by a simple closed polygon). Any doubts that this can be done while keeping the edges straight are dispelled by Fáry's Theorem (1.34); if the reader does not want to accept what was not fully proved in Chapter 1 then he should be content with knowing that V is obtained from a triangulation of a 2-sphere by removing the star of a vertex (by 'punching a hole in the sphere'). Hence:

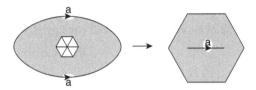

Proposition 9.21. *Let U be a surface with non-empty boundary satisfying $H_1(U;2) = 0$. Then U is isomorphic to a triangulation of the plane region bounded by a simple closed polygon – a disk.* □

Returning to the previous discussion, this shows that a connected maximal non-separating graph K in a closed surface M has the property that the surface V (defined by 9.3) is a triangulation of a disk: 'cutting along K the closed surface M turns into a disk'. If M is represented by a polygonal region in the plane (see the polygonal representations of Chapter 2) such a graph K is given by the boundary polygon.

Representation of homology elements by simple closed polygons

Given a closed surface M, which elements of $H_1(M)$ and $H_1(M;2)$ are the homology classes of 1-cycles given by simple closed polygons which are subcomplexes of M?

Take M to be a torus, so that $H_1(M) \cong \mathbb{Z} \oplus \mathbb{Z}$. Consider the exact sequence (K being a simple closed polygon contained in M)

$$H_1(K) \xrightarrow{i_*} H_1(M) \xrightarrow{j_*} H_1(M,K) \to \widetilde{H}_0(K) = 0.$$

Now $H_1(M,K) \cong H_1(V)$ by Lefschetz duality (9.17) and, V being a disjoint union of surfaces with non-empty boundary, $H_1(V)$ is free abelian (see 9.11(5)). Hence by exactness, $H_1(M)/\operatorname{Im} i_*$ is free abelian. Suppose that the simple closed polygon K represents (when oriented) $p\alpha + q\beta$ where p and q are integers and α and β are free generators for $H_1(M)$. Thus $\operatorname{Im} i_*$ is infinite cyclic with generator $p\alpha + q\beta$, unless $p = q = 0$ in which case $\operatorname{Im} i_* = 0$. In any case $H_1(M)/\operatorname{Im} i_*$ has presentation

$$\left(\begin{matrix} \alpha, \beta \\ p\alpha + q\beta = 0 \end{matrix} \right)$$

so that (compare A.22), $H_1(M)/\operatorname{Im} i_* \cong \mathbb{Z} \oplus \mathbb{Z}_h$ where h is the greatest common divisor of p and q, or $\mathbb{Z} \oplus \mathbb{Z}$ if $p = q = 0$. Since this is free abelian, either $p = q = 0$ or $h = 1$. Hence:

Proposition 9.22. *Let M be a torus and let α, β be free generators for $H_1(M)$. Then an element of $H_1(M)$ representable by a simple closed polygon must be either 0 or $p\alpha + q\beta$ where p and q are coprime. (If $p = 0$ this says $q = 0, 1$ or -1.)* □

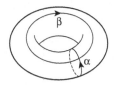
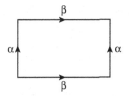

Taking α and β to be the standard generators the reader can probably convince himself informally that, say, 2α cannot be represented by a simple closed polygon: 'it is impossible to draw a simple curve on the surface which goes round twice one way but not at all the other way before joining up'. And given a triangulation there may be elements which are of the form $p\alpha + q\beta$ with p and q coprime but which cannot be represented by simple closed polygons: $n\alpha + \beta$ where n is the number of edges in the triangulation is a pretty safe bet. Nevertheless it is not difficult to believe (and it is true) that any element of the given form can be represented on M or on some barycentric subdivision of M. Thus to this extent the condition of 9.22 is both necessary and sufficient. For example $3\alpha + 2\beta$ can be indicated informally as in the diagram below. The curve ABCDA can be approximated by a graph on a suitably fine barycentric subdivision of any given triangulation.

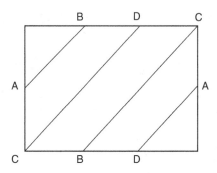

The construction for general coprime p and q can be seen informally as follows. Draw the grid of lines in the plane given by $x = \lambda$ and $y = \mu$ for all integers λ and μ. A torus surface is obtained from the plane by identifying (x, y) with (x', y') whenever $x - x'$ and $y - y'$ are both integers, for the identification means that each (x, y) in the plane is identified with a point in the grid square with corners $(0, 0)$, $(0, 1)$, $(1, 0)$ and $(1, 1)$, and the edges of this grid square are then identified in the usual way to yield a torus. (The identification can be pictured as rolling up the plane to give an infinite cylinder, and then rolling at right-angles to give a torus.) Now draw the straight line from $(0, 0)$ to (q, p). It is a pleasant exercise to check that no two points on this line, except the two end-points, are identified in passing to the torus; hence on the surface the line becomes a simple closed curve which clearly winds q times round one way and p times the other.

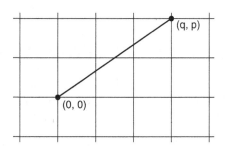

The situation on an arbitrary orientable closed surface is similar. The argument leading to 9.22 yields the following result.

Proposition 9.23. *Let M be an orientable closed surface. Suppose that an element ξ of $H_1(M)$ is the homology class of a cycle given by a simple closed polygon. Then ξ is indivisible, i.e. not a multiple of any element of $H_1(M)$ besides $\pm \xi$, or else $\xi = 0$.* □

In fact the exact sequence used to prove 9.22 shows that $H_1(M)/\langle\xi\rangle$ is free abelian, $\langle\xi\rangle$ being the subgroup of $H_1(M)$ generated by ξ, and the result follows easily from this. It is also true (see A.34(4)) that if ξ is indivisible or zero then $H_1(M)/\langle\xi\rangle$ is free abelian, and indeed it can be shown that such ξ are always representable on M or some barycentric subdivision of M by simple closed polygons.[†] Granted this, it follows that every element of $H_1(M;2)$ is representable on M or some barycentric subdivision of M by a simple closed polygon.

Note that Lefschetz duality gives no information when M is non-orientable. The reader may like to consider informally such questions as: Can every element of $H_1(M;2)$, where M is non-orientable, be represented on M or some barycentric subdivision of M by a simple closed polygon? (The answer is in fact 'yes', and the reader is left with no hints beyond the diagram in 9.25(2), where the arrows should be removed from the dotted lines. (Next, try $a_1 + a_2$!))

Orientation preserving and reversing loops

Let K be a simple closed polygon which is a subcomplex of a closed surface M, and consider the regular neighbourhood N of K in M (9.3). Certainly N is connected, since $N \searrow K''$ by 9.5; hence N is a surface (9.15) with non-empty boundary $N \cap V$. Consider the corresponding closed surface $\overline{N}$ (9.11(4)) obtained by adding a cone on each boundary component of N. Thus if N has n boundary components, $\chi(\overline{N}) = \chi(N) + n = n$ since $\chi(N) = \chi(K'') = 0$, and since $\chi(\overline{N}) \leqslant 2$ (2.14) this implies $n = 1$ or 2. Correspondingly $\chi(\overline{N}) = 1$ or 2 so $\overline{N}$ is a projective plane or sphere. The N we started with is therefore a Möbius band or cylinder (compare 9.11(4)).

Definition 9.24. The simple closed polygon K in the above discussion – or the loop on M which it determines – is called *orientation preserving* or *reversing* according as the regular neighbourhood is a cylinder or Möbius band.

The geometrical meaning of this definition is that a small oriented circle lying in the surface, whose centre traverses the loop once, returns with its orientation preserved or reversed according as N is a cylinder or Möbius band.

Examples 9.25. (1) If M is orientable, so are N and $\overline{N}$; hence every loop on an orientable surface is orientation preserving.

(2) Let M be non-orientable. Then $H_1(M) \cong \mathbb{Z}^{k-1} \oplus \mathbb{Z}_2$ where k is the genus of M (6.13). The diagram shows (in the case $k = 3$) how to represent the unique

---

[†] This result was pointed out to me by H. R. Morton; see Poincaré (1904) p. 70, or Schafer (1976).

element of order 2 in $H_1(M)$ by a simple closed polygon – it may be necessary to subdivide M in order to get a genuine polygon which is a subcomplex of M. (To see why it represents $\{a_1+a_2+a_3\}$, note that $PQ+QR+RP \sim PO+OQ+QO+OR+RO+OP = OP+PO+OQ+QO+OR+RO = a_1+a_2+a_3$.)

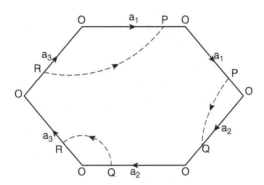

Suppose, then, that K is any simple closed polygon in M which gives a 1-cycle representing the unique element of order 2 in $H_1(M)$. Is K orientation preserving or reversing? Now reducing mod 2 it follows that K gives a 1-cycle mod 2 homologous (mod 2) to $a_1 + \cdots + a_k$ and therefore not representing $0 \in H_1(M;2)$ (see 8.5). Hence K does not separate M (9.19): in the usual notation V is connected. Also from the homology sequence of (M,K) it follows that $H_2(M,K) \cong \mathbb{Z}$ (note the integer coefficients here), so that $\mathbb{Z} \cong H_2(M'',K'') \cong H_2(M'',N) \cong H_2(V,\partial V)$ (see 9.6, 9.7 which do not depend on M being orientable). Since V is a single surface this shows that it is orientable (see 9.12): to put it more picturesquely, *cutting M along K it turns into an orientable surface*.

Since $M'' = N \cup V$, and $N \cap V$ consists of simple closed polygons, which have Euler characteristic 0, we have $2 - k = \chi(M) = \chi(M'') = \chi(N) + \chi(V) = \chi(V)$ since $\chi(N) = \chi(K) = 0$. Further $\chi(\overline{V}) = \chi(V) + n$ (n = number of boundary components of N, or equally well of V) and $\chi(\overline{V})$ is even since $\overline{V}$ is orientable. Hence $2 - k + n$ = even number, and therefore n is even if and only if k is even: K is orientation preserving if and only if k is even.

In particular for a Klein bottle $k = 2$, and the loop is orientation preserving in this case. The surface V is a cylinder – the Klein bottle becomes a cylinder when cut along K.

(3X) In the notation of (2), show that if K gives a cycle representing any non-zero element of $H_1(M)$ besides the one of order 2 then V is a non-orientable surface.

(4X) Show that if K gives a cycle which is homologous to zero (mod 2) on M, then K is always orientation preserving. (Recall that such a K separates M, so that V has two components.)

(5X) Suppose that K_1 and K_2 are disjoint simple closed polygons which are subcomplexes of some triangulation M of a torus. It is very plausible that they cannot give cycles which freely generate $H_1(M)$: such cycles 'ought to intersect'. To prove this, note first that by (1), the regular neighbourhood of $K = K_1 \cup K_2$ in M has four boundary components. Supposing that K_1 and K_2 do give cycles representing free generators for $H_1(M)$ it follows from the homology sequence of (M, K) and Lefschetz duality that, in the usual notation, V is connected. Since $\chi(V) = \chi(M) = 0$ (compare the last paragraph but one of (2) above), we have $\chi(\overline{V}) = \chi(V) + 4 = 4$, which is impossible for a closed surface $\overline{V}$.

(6X) Suppose that K is a simple closed polygon (subcomplex of a closed surface M) and that K is a boundary component of some surface S contained in M. Then K is orientation preserving. (For, writing N for a regular neighbourhood of K in M, ∂N will have one component in S'' and one outside S''.) Thus, for example, the simple closed polygons bounding the shaded regions in the Klein bottle and projective plane of 2.2(4) are orientation preserving.

Now suppose that K and L are disjoint simple closed polygons in M. Let a and b be the corresponding 1-cycles mod 2 and suppose that $a \sim b$ (mod 2). Then $a + b = \partial c$ for some $c \in C_2(M; 2)$ and it is not hard to check that the 2-simplexes of M occurring in c with coefficient 1, together with their edges and vertices, form a surface with boundary $K \cup L$. Hence, by the above result, K and L are both orientation preserving.

One consequence of this is that if two simple closed polygons are orientation reversing and represent the same element of $H_1(M; 2)$ then they must intersect. The reader should examine some simple closed curves on a Klein bottle to convince himself that this result is plausible.

★ (7XX) The concept of orientation preservation or reversal depends only on the homology class mod 2 of the polygon. More precisely, let K and L be simple closed polygons, subcomplexes of a closed surface M, and let the corresponding 1-cycles mod 2 be a and b. *If $a \sim b$ (mod 2) then K and L are both orientation preserving or both orientation reversing.* Here is a very brief and informal outline of a possible argument for this. All homology is mod 2 in this example.

Suppose that, as above, $a \sim b$. First we want to change K and L so that they meet only in points at which they cross. This is called moving K and L to 'general position'. By subdividing M barycentrically enough times we can move K and L to general position by arbitrarily small moves, affecting neither the homology classes they represent nor their properties of orientation preservation or reversal. (It would not be difficult to give an elementary proof of this; the proof would however be full of rather intricate detail, and we shall assume the result here.) It is therefore enough to prove the above italicized result when K and L are in general position.

Now $a \sim b$ so $a + b = \partial c$ for some $c \in C_2(M; 2)$. Consider the part of M, U say, covered by the 2-simplexes of M with coefficient 1 in c. At a crossing of K and L vertically opposite segments belong to U, as in the left-hand diagram below, where U is shaded. For each crossing we perform the operation shown (perhaps subdividing M first) turning U into U' and $K \cup L$ into B say. Then U' is a collection of disjoint surfaces with total boundary B. Each component of B will be orientation preserving, by the first part of (6) above. Let N be a regular neighbourhood of B; we are interested in the number of boundary components of N. By the property of B just mentioned this number will certainly be even.

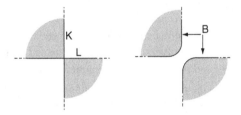

Now suppose that K is orientation preserving and L is orientation reversing. Then a regular neighbourhood of K has boundary with two components and a regular neighbourhood of L has connected boundary. Of course these neighbourhoods will intersect near the points of $K \cap L$ but let us imagine instead that one passes 'under' the other. Then we have two surfaces (the now disjoint neighbourhoods of K and L) with $1 + 2 = 3$ boundary components and a sequence of operations corresponding to that pictured above turns the surface into N. Then each operation changes the number of boundary components by an even number, possibly 0. Hence N has $3 +$ even number of boundary components, which will be odd. This contradiction proves the result. □

Here is one consequence. On a closed surface kP take the basis $\{a_1\}, \ldots, \{a_k\}$ for $H_1(kP; 2)$ (compare 8.5(2), 2.8). Note that $a_1, \ldots, a_k$ are all orientation reversing; in fact we have the following. A simple closed polygon in kP is orientation reversing if and only if it represents $\lambda_1\{a_1\} + \cdots + \lambda_k\{a_k\} \in H_1(kP; 2)$ with $\lambda_1 + \cdots + \lambda_k = 1$ (mod 2). Compare (2) above. ★

(8XX) Show that, when K is a graph in a closed surface M, the number of orientable components of V is the rank of $H_2(M, K)$. (Compare part of (2) above.) Now let M be a projective plane and denote by n the rank of

$H_1(K)$ (or equally well the dimension of $H_1(K;2)$). Show that the number of orientable components of V is always n and that there is never more than one non-orientable component of V.

More generally, show that if $M = kP$ then there are never more than k non-orientable components of V. [Hint. First show that the number of orientable components is $\geqslant n + 1 - k$ and then that the total number of components is $\leqslant n + 1$.]

A generalization of Euler's formula

Theorem 9.26. *Let K be a non-empty graph contained in a closed surface M, and let $V_1, \ldots, V_r$ be the components of V (in the usual notation). Then*

$$\alpha_0(K) - \alpha_1(K) + r = \chi(M) + \sum_{i=1}^{r} \hat{\beta}_1(V_i)$$

where, as usual, $\hat{\beta}_1$ denotes the dimension of a first homology group with $\mathbb{Z}_2$ coefficients.

Proof Now $M'' = V \cup N$ and $V \cap N$ is a union of simple closed polygons, so $\chi(M) = \chi(M'') = \chi(V) + \chi(N) = \chi(V) + \chi(K)$ since $N \searrow K''$. The result now follows because $\chi(V_i) = 1 - \hat{\beta}_1(V_i)$, V_i being a surface with non-empty boundary (since $K \neq \emptyset$) for $i = 1, \ldots, r$. □

Corollary 9.27. *('Euler's inequality') In the notation of 9.26,*

$$\alpha_0(K) - \alpha_1(K) + r \geqslant \chi(M)$$

with equality if and only if each V_i is a triangulation of a disk.

Proof This follows from 9.26 and 9.21. □

This result generalizes Euler's formula (1.27), which refers to the case when M is a sphere, or equally well to the case of a planar graph (see the discussion following 9.2). When K is connected and M is a sphere it follow easily from the homology sequence (mod 2) of (M, K) and Lefschetz duality that $H_1(V;2) = 0$.

Examples 9.28. (1) Let K be the 1-skeleton of M. Then equality holds in 9.27, which is just the definition of $\chi(M)$.

(2X) Suppose that there is some way of realizing a graph K as a subcomplex of a closed surface so that all the components of V are triangulations of disks (i.e. 'all regions are open disks' – 9.16(2)). Show from 9.27 that the number of regions is minimal, i.e. ⩽ the number of regions in any other way of realizing K as a subcomplex of M. (Compare the diagrams after 9.2.)

★ (3X) Suppose that K is a connected non-empty graph contained in a closed surface M and that some component U of V (usual meaning for V) is not a triangulation of a disk. Let ∂U be the union of $p > 0$ simple closed polygons $S_1, \ldots, S_p$. Remove U from M'' and replace it by disjoint cones on $S_1, \ldots, S_p$. The resulting simplicial complex M_1 is connected (since K is connected) and is in fact a closed surface with K'' as a subcomplex. Also $\chi(M_1) = \chi(M) + p - \chi(U)$. Now $\chi(U) \leqslant 1$ (since $\partial U \neq \emptyset$) and indeed $\chi(U) < 1$ since U is not a triangulation of a disk (compare 9.21). Hence $\chi(M_1) > \chi(M)$. This shows that K'' is realized in a closed surface of higher Euler characteristic than M. According to a general theorem on graphs in surfaces this implies that K can also be realized in a closed surface of Euler characteristic $> \chi(M)$ (compare (4) below). Hence we have the following result.

Suppose that the connected graph K can be realized as a subcomplex of a closed surface M but not as a subcomplex of any closed surface N with $\chi(N) > \chi(M)$. Then, however K is realized in M, all regions will be open disks (i.e. all components of V will be triangulated disks; compare 9.16(2).)

Since M orientable implies M_1 orientable we also have:

Suppose that the connected graph K can be realized in an orientable closed surface M but not in any orientable closed surface N of lower genus than M. Then the same conclusion holds. (Is this true with 'non-orientable' replacing 'orientable'?)

As an example of this result, the complete graphs G_5, G_6, G_7 can be realized in a torus, but not in a sphere (compare 1.32). Hence no matter how they are realized in a torus, the regions will be open disks. ★

★ (4X) Here is a sketch of a proof of the following result.

If some subdivision K_1 of a graph K is a subcomplex of a closed surface M_1 then K is a subcomplex of an equivalent closed surface.

This resembles a version of Fáry's theorem (1.34), but is much easier to prove. It is enough to consider the case when the subdivision K_1 is obtained by adding one vertex v on an edge (uw) of K. We consider the link of v in M_1, drawn in the left-hand picture.

A generalization of Euler's formula

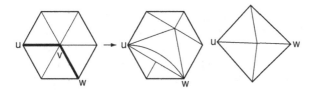

Suppose for the moment that u and w are not adjacent in the link. Now (uv) and (vw) belong to K_1 but the other edges radiating from v cannot belong to K_1 since $v \notin K$. Transform as shown in the middle picture, removing v and introducing two new vertices. It is possible that the transformed M_1 now contains two 1-simplexes (uw) – the one drawn curved, and, possibly, one already present in M_1. In that case transform the one already present (which cannot belong to K_1) as shown in the right-hand diagram. This approach, followed by the same transformation, also deals with the case when u and w are adjacent on the link of v. Finally straighten everything by appeal to the general realization theorem of 3.19. The result is K as a subcomplex of a closed surface which is clearly equivalent to M_1. (It is even more clearly p.l. equivalent to M_1 – see 5.9.)

Why does this approach not prove the corresponding result for planar graphs? ★

(5X) The converse result to that in (3) – 'all regions are open disks implies that the graph cannot be realized in a closed surface of higher Euler characteristic' – is false. Find a counterexample.

(6X) It has already been noted that for planar graphs the removal of an edge decreases the number of regions by one if and only if the edge belongs to a loop (see 1.26). Show that for a graph in an arbitrary closed surface, removing an edge from the graph decreases the number of regions (by one) if and only if the edge belongs to a loop which is homologous to zero (mod 2).

★ (7X) One feature of the inequality 9.27 is that both sides are invariant under subdivision of M. (Compare 2.16(8).) We might say that 9.27 expresses something intrinsic about the situation, not dependent on the particular triangulation. An inequality of a different stamp is, in the same notation, $3r \leqslant 2\alpha_1(K)$. Here, the right-hand side increases with subdivision but the left-hand side remains fixed. Other peculiarities of this inequality are its extreme weakness ($2\alpha_1(K) - 3r$ tends to be not merely positive but very large), which makes it difficult to prove by induction, and the anomaly that it is false when $\alpha_1 = 0$ or 1.

It is very plausible that each region has at least three edges of K enclosing it (at any rate when K has at least three edges), and certainly each edge of K is adjacent to at most two regions. So the sum of the numbers of edges of K

adjacent to the various regions should be $\geqslant 3r$ and $\leqslant 2\alpha_1$, whence $3r \leqslant 2\alpha_1$. The argument below makes this more precise.

Consider a 1-simplex e of K. Running along the two sides are boundary components of surfaces V_1 and V_2 which are components of V (possible $V_1 = V_2$). Call e *adjacent* to these boundary components and also to V_1 and V_2.

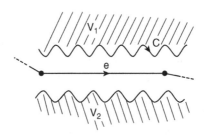

Suppose $V_1 \neq V_2$. Traverse the boundary component C of V_1 adjacent to e in a particular direction, noting the succession of edges of K which are adjacent and giving them as they occur the direction of C. (These edges are the ones immediately, say, 'on the right' of C; note that by 9.25(6) C is orientation preserving!) Since $V_1 \neq V_2$ we eventually return to e and assign it the same direction as before. Hence the edges of K adjacent to C contain a loop. (If $V_1 = V_2$ this result may or may not be true: the reader should look for examples of both events.)

Use the result just given and induction of $\alpha_1(K)$ to prove that, provided K is not a disjoint union of trees, every component of V has a loop of edges adjacent to it. In particular, provided K is not a disjoint union of trees, every component of V has at least three edges of K adjacent to it. What if K is a disjoint union of trees? (Note that in that case there is only one region.) Now use the fact that each edge of K is adjacent to at most two components of V to prove that $3r \leqslant 2\alpha_1(K)$ provided $\alpha_1(K) \geqslant 2$.

If K satisfies $3r = 2\alpha_1(K)$ then it follows that every component of V has precisely three edges of K adjacent to it, forming a loop. Hence, provided K has no isolated vertices, every region is bounded by a triangle. This proves something stated following 2.22: in the notation used there, every region into which G_λ divides the surface is a disk, by (3) above; hence by 9.27 we have the equality $\alpha_0(G_\lambda) - \alpha_1(G_\lambda) + r = \chi$. Using this, and $\alpha_0 = \frac{1}{2}(7 + \sqrt{(49 - 24\chi)})$, $\alpha_1 = \frac{1}{2}\alpha_0(\alpha_0 - 1)$ it follows that $3r = 2\alpha_1$, so that G divides the surface into triangles.

The above analysis shows that a graph G in a surface S divides the surface into triangles if and only if, in the usual notation, $\alpha_0 - \alpha_1 + r = \chi$ and $3r = 2\alpha_1$. It is worth noting that these two conditions can be combined into a single condition, namely $3\alpha_0 - \alpha_1 = 3\chi$. For we always have $\alpha_0 - \alpha_1 + r \geqslant \chi$ and $2\alpha_1 \geqslant 3r$, so that $3\alpha_0 - \alpha_1 \geqslant 3\alpha_0 - 3\alpha_1 + 3r \geqslant 3\chi$. Hence $3\alpha_0 - \alpha_1 = 3\chi$ if and only if both the preceding inequalities are equalities. Compare the book by Ringel, listed in the References. ★

★ (8X) The result $3r \leqslant 2\alpha_1$ of (7), together with 'Euler's inequality', can be used to give a proof, essentially Heawood's, of Heawood's Theorem, quoted at the end of Chapter 2: Every graph in a closed surface, other than the sphere, of Euler characteristic χ can be coloured with $\mathcal{H}(\chi)$ colours, where

$$\mathcal{H}(\chi) = \left\lfloor \tfrac{1}{2}\left(7 + \sqrt{49 - 24\chi}\right) \right\rfloor.$$

Here are some hints. Prove that the average of the orders of the vertices of K is $2\alpha_1(K)/\alpha_0(K)$ and that this is $\leqslant 6 - (6\chi(M)/\alpha_0(K))$. From this it follows that there is at least one vertex where $\leqslant 6 - (6\chi(M)/\alpha_0(K))$ edges meet.

Suppose now that $\chi(M) \leqslant 0$, and prove the following claim by induction on $\alpha_0(K)$: If c is a positive integer satisfying the inequality

$$6 - (6\chi(M)/c) \leqslant c - 1$$

then any graph in M can be coloured with c colours. (Note that the conclusion is obvious if $\alpha_0(K) \leqslant c$, so start the induction at c.) Now prove that the smallest c satisfying the hypothesis of the claim is $\mathcal{H}(\chi)$.

Finally consider $\chi(M) = 1$ (projective plane). When $\chi(M) = 2$ the method only gives the uninteresting result that every graph in the sphere can be coloured with six colours. The stronger result with four in place of six is known to be true – this is the Four Colour Theorem; see for example the textbooks of Wilson or Gross & Yellen listed in the References and page 66 above. ★

★ Brussels Sprouts

As a final example of graphs in surfaces we shall consider an amusing game due to J. H. Conway, and called (by him) *Brussels Sprouts*. (One case of this game, together with another more serious game called Sprouts, was described by Martin Gardner in the July 1967 issue of *Scientific American*; this article together with a commentary appears in the collection *Mathematical Carnival* listed in the References. See also the textbook of Firby & Gardiner listed in the References. Brussels Sprouts is a pencil and paper, or pencil and surface, game which involves drawing 'graphs' with, probably, curved edges in a curved surface or in some plane representation of one. Thus we shall not deal with explicit triangulations, but hope instead that the reader is prepared to believe what was claimed in Chapter 2, namely that a 'curved graph' can be drawn in a 'curved surface' if and only if the corresponding straight graph can be realized as a subcomplex of a (triangulated) closed surface in the appropriate equivalence class. (The sceptical reader can, of course, avoid this by always playing the game on a (preferably very fine) triangulation!)

The game starts with $n \geqslant 1$ crosses in a closed surface, and there are two players who move alternately. Each cross provides 4 arms or 'sprouts' and a move is to join two sprouts with a simple arc (one that does not intersect itself) and draw a small crossing line somewhere on the arc. Two sample first moves with $n = 2$ are drawn below.

The two sprouts at the ends of the arc are not available for future moves, but two new ones have appeared, at the ends of the crossing line. No arc is allowed to intersect any of the rest of the figure except at the beginning and end of the arc, which must be the end-points of previously unused sprouts. Eventually, as we shall see, no more moves are possible; *the last person to move is the winner.*

At every stage of the game we have a curved graph in a closed surface M and after p moves ($p \geqslant 0$) the graph, K say, has $\alpha_0(K) = 5n + 3p$ and $\alpha_1(K) = 4n + 4p$. Furthermore since a sprout is added on each side of every arc drawn during the game it is clear that at least one sprout must point into every region into which K divides M. Since the number of sprouts is constant at $4n$ we have that r, the number of regions, is $\leqslant 4n$. Applying Euler's inequality (9.27), this gives $p \leqslant 5n - \chi(M)$. Hence:

Proposition 9.29. *The game of Brussels Sprouts on a closed surface of Euler characteristic χ must end after at most $5n - \chi(M)$ moves.* □

Presumably the game can always be played so that it doesn't end before $5n - \chi(M)$ moves. Note that at the end of such a 'maximal' game all the regions into which K divides M would be disks, by 9.27.

Now let us look more closely at the position when the game has finished, which happens, say, after m moves. At the end there will be exactly $4n$ regions, with a sprout pointing into each one (otherwise a further move is possible). Taking a regular neighbourhood N of K and complementary union of surfaces V as usual, it follows that the boundary of each component of V is connected. (For if a component V_1 of V had disconnected boundary, the region corresponding to V_1 would have a sprout pointing into it near each component of ∂V_1, and a further move would be possible.)

Suppose now that M is orientable. Then each component of V is an orientable surface with connected boundary. By adding a cone on this boundary to obtain an orientable closed surface, which will have even Euler characteristic, it follows that $\chi(V_i)$ is odd for each component V_i of V; consequently $\hat{\beta}_1(V_i) = \dim H_1(V_i; 2)$ is even. Hence from 9.26,

$$\alpha_0(K) - \alpha_1(K) + r - \chi(M) \text{ is even}$$

which gives, $\chi(M)$ being even,

$$n - m \quad \text{is even.}$$

This proves the following result, which says that playing Brussels Sprouts on an orientable surface, the first player is is doomed to succeed or fail from the start.

Theorem 9.30. *For the game of Brussels Sprouts on an orientable closed surface, the first player wins (i.e. m is odd) if and only if the number of crosses to begin with is odd.* □

It is not difficult to prove that

$$5n - 2 \leqslant m \leqslant 5n - \chi(M)$$

whether M is orientable or not. Presumably when M is orientable m can take any value congruent to n (mod 2) between these bounds. If M is a sphere, $\chi(M) = 2$ so *the actual number of moves is determined by n*: this is the game played in the plane.

The case when M is non-orientable is more tricky, and only a few remarks on the subject will be made here.

Proposition 9.31. *On a non-orientable surface M with standard form $a_1 a_1 \ldots a_k a_k$ (sphere with k crosscaps), starting with n crosses, there is a strategy for the first player to win unless n and k are both even.*

Hints for proof Letting m be the number of moves in a game, we have, from 9.26

$$5n - m = 2 - k + \sum \hat{\beta}_1(V_i).$$

(Compare the argument leading to 9.30.)

Case I k even, n odd. Then the first player wins if and only if $\sum \hat{\beta}_1(V_i)$ is even. He can ensure this by drawing a loop on M which represents the unique element of order 2 in $H_1(M; 2)$; the game is then essentially on an orientable surface (compare 9.25(2)), and this implies that each $\hat{\beta}_1(V_i)$ is even.

Case II k odd, n odd. The first player requires $\sum \hat{\beta}_1(V_i)$ to be odd, and he can ensure this by enclosing all the crosses in a disk, leaving one sprout pointing out of it:

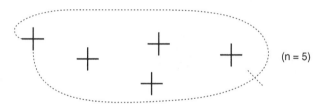

(Then all V_i but one are orientable.)

Case III k odd, n even. As in Case I. □

The remaining case, k and n both even, is more difficult and I claim that there is a strategy for the second player to win. Take $k = 2$ (Klein bottle); then in the notation of 9.2 (and letting m be the total number of moves),

$$m = 5n + d - 2.$$

Note that $0 \leq d \leq 2$ and that the first player wins (n being even) if and only if $d = 1$. Thus I claim that by playing suitably the second player can prevent the game from ending with $d = 1$.

For general n and k (both even), the reader may like to prove the following. The first player's first move must be a *closed* curve (i.e. both ends must be at sprouts on the same cross) which *separates* the closed surface into two *non-orientable* surfaces in *both* of which further play is possible. (There are four assertions here, given by the italicized words; the word 'must' means that if the first player fails to do any of these four things then the second player can ensure a win by his first move. If the first player does do these and continues to play rationally then the second player must keep on his toes at every subsequent move in order to win.) ★

Appendix: abelian groups

This appendix presents an account of the elementary theory of abelian groups which is heavily biased towards quotient groups, exact sequences and presentations, these being the main ideas from the theory which are used in the body of the book. In fact scarcely anything is said which is not strictly relevant to the material of the book, the only exceptions being a few partially worked examples. There are accounts of the general theory of abelian groups in many texts on algebra or group theory, for example in the books of Lang and Mac Lane & Birkhoff listed in the References. These books give full details of the classification theorem for finitely generated abelian groups, stated but not proved in A.27. However, every result from abelian group theory which is actually used in the book is proved in this appendix.

One piece of notation: here and in the text the value of a map f at an element x in its domain is denoted by fx rather than $f(x)$, unless there is a special need for bracketing as in $f(x+y)$.

Basic definitions

A.1. Abelian group An *abelian group* is a set A, together with a binary operation $+$ defined on A and satisfying the following four axioms:

(1) For all a, b, c in A, $a + (b + c) = (a + b) + c$.
(2) There exists $0 \in A$ such that, for all $a \in A$,

$$a + 0 = 0 + a = a.$$

(3) For each $a \in A$ there exists $b \in A$ such that

$$a + b = b + a = 0.$$

(4) For all a, b in A, $a + b = b + a$.

It follows from these axioms that the 0 of (2) is unique and that, in (3), b is uniquely determined by a. We write $-a$ for this unique b and abbreviate $c + (-a)$ to $c - a$ for any $c \in A$.

Throughout this appendix the letters A, B, C will, unless otherwise stated, stand for arbitrary abelian groups. As is customary, we refer to a group using just the name of its underlying set.

A.2. Subgroup A *subgroup* of A is a subset of A which is also an abelian group under the same binary operation.

It follows that $B \subset A$ is a subgroup of A if and only if B is non-empty and, for all $a, b \in B$, we have $a - b \in B$.

A.3. Trivial group Any group with precisely one element is denoted 0 and called a *trivial* or *zero group*.

A.4. Let $a \in A$ and $n \in \mathbb{Z}$. By na we mean 0 if $n = 0$; $a + a + \cdots + a$ (n summands) if $n > 0$; and $(-a) + \cdots + (-a)$ ($-n$ summands) if $n < 0$. thus $(-n)a = -(na)$ for any n.

A.5. Order Let $a \in A$. If, for all $n \in \mathbb{Z}$, $n \neq 0$, we have $na \neq 0$ then a is said to have *infinite order*. Otherwise, the *order* of a is the smallest integer $n > 0$ such that $na = 0$. Thus a has order 1 if and only if $a = 0$ and, for any a, a and $-a$ have the same order, or both have infinite order. The *order* $|A|$ *of a group* A is simply the number of elements in A.

A.6. Cyclic group A is called *cyclic* if there exists $a \in A$ such that any $b \in A$ is of the form na for some $n \in \mathbb{Z}$. Such an a is called a *generator* of A. Note that $-a$ is also a generator.

Examples: (1) The integers $\mathbb{Z}$ under addition is a cyclic group with generator 1 (or, alternatively, -1), which has infinite order.

(2) The set $\{0, 1, \ldots, k-1\}$ ($k \geq 1$) under addition modulo k is a cyclic group $\mathbb{Z}_k$ with generator 1 (or, in fact, any element coprime to k), which has order k. Note that $\mathbb{Z}_1 = 0$.

Note that we use the notation $\mathbb{Z}_k$ for a cyclic group of order k. Other common notations are $\mathbb{Z}/k$ or $\mathbb{Z}/k\mathbb{Z}$.

(3) For any A and $a \in A$ the subset $\{na : n \in \mathbb{Z}\}$ is a subgroup of A which is cyclic, a being a generator. It is called the subgroup *generated by* a and has order equal to the order of a (infinite order if a has infinite order).

A.7. Homomorphism A *homomorphism* f from A to B is a map $f : A \to B$ such that, for all $a_1, a_2 \in A, f(a_1 + a_2) = fa_1 + fa_2$. (Here the binary operations in both groups are denoted +.) This implies $f0 = 0$ and $f(-a) = -(fa)$ for any $a \in A$.

The *domain* of f is A and the *target* of f is B.

Examples: (1) The *trivial* homomorphism $f : A \to B$ is defined by $fa = 0$ for all $a \in A$; we write $f = 0$. (Thus '0' now has three meanings besides its customary one of 'the integer zero'.) Any homomorphism $0 \to A$ or $A \to 0$ is trivial.

(2) The identity map $1_A : A \to A$ is a homomorphism. If A is a subgroup of B then the inclusion $i : A \to B$ defined by $ia = a$ for all $a \in A$ is a homomorphism.

(3) If $f : A \to B$ and $g : B \to C$ are homomorphisms then so is the composite $g \circ f : A \to C$, defined by $(g \circ f)a = g(fa)$ for all $a \in A$.

A.8. Kernel, image, monic, epic, isomorphism Let $f : A \to B$ be a homomorphism. We define

$$\text{Kernel of } f = \text{Ker} f = \{a \in A : fa = 0\};$$
$$\text{Image of } f = \text{Im} f = \{b \in B : b = fa \text{ for some } a \in A\}.$$

These are subgroups of A and of B respectively.

We call f *monic* or a *monomorphism* or *injective* if $\text{Ker} f = 0$.
We call f *epic* or an *epimorphism* or *surjective* if $\text{Im} f = B$.
We call f an *isomorphism* if both $\text{Ker} f = 0$ and $\text{Im} f = B$.

An isomorphism is bijective and conversely any bijective homomorphism is an isomorphism. If f is an isomorphism then so is its inverse f^{-1}. Groups A and B are called *isomorphic*, denoted $A \cong B$, if there exists an isomorphism $f : A \to B$. The relation $\cong$ is an equivalence relation. Note that, in accordance with A.3, we write $A = 0$ rather than $A \cong 0$.

Example: any cyclic group is isomorphic either to $\mathbb{Z}_k$ for some $k \geqslant 1$ or to $\mathbb{Z}$ (see A.6).

Finitely generated (f.g.) and free abelian groups

A.9. Finitely generated abelian group Suppose that $a_1, \ldots, a_n$ are elements of A with the property that every $a \in A$ can be expressed in the form

$$a = \lambda_1 a_1 + \lambda_2 a_2 + \cdots + \lambda_n a_n$$

with $\lambda_1, \ldots, \lambda_n$ integers (compare A.4). Then A is called *finitely generated* *(f.g.)* with *generators* $a_1, \ldots, a_n$. The expression for a is called an *integral linear combination* of $a_1, \ldots, a_n$.

A.10. Free f.g. abelian group Suppose that A is f.g. with generators $a_1, \ldots, a_n$. Suppose in addition that, for all integers $\lambda_1, \ldots, \lambda_n$,

$$\lambda_1 a_1 + \cdots + \lambda_n a_n = 0 \text{ implies that } \lambda_1 = \cdots = \lambda_n = 0.$$

Then A is called a *free f.g. abelian group* and $a_1, \ldots, a_n$ are called *free generators* or a *basis* for A. (Since all groups considered in the text are f.g. we usually refer there to *free abelian groups*, omitting 'f.g.' from the description.) By convention 0 is free with empty basis. It is proved in A.30 that any two bases of the same free f.g. abelian group contain the same number of elements.

Thus A is free and f.g. with basis $a_1, \ldots, a_n$ if and only if every $a \in A$ can be expressed uniquely as an integral linear combination of $a_1, \ldots, a_n$. The free f.g abelian groups are analogous to finite dimensional vector spaces, with the field over which the vector spaces are defined being replaced by the group $\mathbb{Z}$ of integers under addition. Some results about free f.g. abelian groups are in fact derived from the corresponding results about vector spaces at the end of this appendix.

Examples: (1) $\mathbb{Z}$ is f.g.; for generators we can take, for example, the two elements 2, 3, or alternatively just 1, or alternatively -1. The generator 1 is a basis; so is -1; but the two elements 2, 3 are not.

(2) $\mathbb{Z}_k$ is f.g. but not free, since for any $a \in \mathbb{Z}_k$ we have $ka = 0$.

(3) Let $\mathbb{Z}^n$ be the group of n-tuples of integers $(\lambda_1, \ldots, \lambda_n)$ under the operation

$$(\lambda_1, \ldots, \lambda_n) + (\mu_1, \ldots, \mu_n) = (\lambda_1 + \mu_1, \ldots, \lambda_n + \mu_n).$$

Then $\mathbb{Z}^n$ is a free f.g. abelian group with basis

$$(1, 0, \ldots, 0), (0, 1, 0, \ldots, 0), \ldots, (0, 0, \ldots, 0, 1).$$

This is often called the *standard basis* for $\mathbb{Z}^n$.

★(4) The group $\mathbb{Q}$ of rational numbers under addition is not f.g. For let $a_1 = p_1/q_1, \ldots, a_n = p_n/q_n$ be rational numbers, with the p_i and q_i integers. Then any integral linear combination of $a_1, \ldots, a_n$, when reduced to its lowest terms, must have denominator a factor of $q_1 \ldots q_n$. Since there exist rational numbers without this property, e.g. $1/2q_1 \ldots q_n$, the given rational numbers do not generate $\mathbb{Q}$. ★

A.11. Free abelian group on a set The most important example for our purposes, arising in the text when defining chain groups, is as follows. Let $S = \{x_1, \ldots, x_n\}$ where the x's are distinct. Let FS be the set of expressions

$$\sum \lambda_i x_i = \lambda_1 x_1 + \cdots + \lambda_n x_n$$

where $\lambda_1, \ldots, \lambda_n$ are integers. Two elements of FS, $\sum \lambda_i x_i$ and $\sum \mu_i x_i$ say, are defined to be equal if and only if $\lambda_i = \mu_i$ for $i = 1, \ldots, n$. Addition is defined by

$$\sum \lambda_i x_i + \sum \mu_i x_i = \sum (\lambda_i + \mu_i) x_i.$$

Then FS is an abelian group; in fact it is free and f.g. with basis $1x_1 + 0x_2 + \cdots + 0x_n, 0x_1 + 1x_2 + 0x_3 + \cdots + 0x_n, \ldots, 0x_1 + \cdots + 0x_{n-1} + 1x_n$. If we agree to omit terms with coefficient zero and to write $1x_i$ as x_i then this basis becomes just $x_1, \ldots, x_n$. We call FS the *free abelian group on* S. (Note that FS depends, strictly speaking, on the ordering of S. This dependence can be eliminated by disregarding the order in which the terms of $\sum \lambda_i x_i$ are written.) Conventionally $F\emptyset = 0$. Notice that $FS \cong \mathbb{Z}^n$, an isomorphism being given by $\lambda_1 x_1 + \cdots + \lambda_n x_n \to (\lambda_1, \ldots, \lambda_n)$.

Proposition A.12. *Let A be free and f.g. with basis $a_1, \ldots, a_n$ and let B be an abelian group. Given elements $b_1, \ldots, b_n$ in B there is a unique homomorphism $f : A \to B$ with $fa_i = b_i$ for $i = 1, \ldots, n$.*

Proof The conditions on f imply that $f(\lambda_1 a_1 + \cdots + \lambda_n a_n) = \lambda_1 b_1 + \cdots + \lambda_n b_n$ for any integers $\lambda_1, \ldots, \lambda_n$. It is straightforward to check that, A being free with basis $a_1, \ldots, a_n$, this does define a homomorphism. □

For example, the boundary homomorphism $\partial_p : C_p(K) \to C_{p-1}(K)$ (see 1.19, 4.3) is completely specified by its values on the p-simplexes of K. In this example both groups are free and f.g.

Quotient groups

The definition of homology groups (4.8) depends on the idea of quotient group. Given a subgroup A of an abelian group B the quotient group B/A is a device for 'ignoring' or 'making zero' the elements of A. In the text A is the boundary subgroup of a cycle group B. Both of these are free abelian, but the following definition does not assume this.

A.13. Cosets and quotient groups Let A be a subgroup of B. For each $b \in B$ the subset

$$\bar{b} = b + A = \{b + a : a \in A\}$$

is called a *coset of A in B*. (From Chapter 4 onwards we use the standard, but rather unfortunate, notation $\{b\}$ for $b + A$. It would be too confusing to use it here.) The set of cosets of A in B is denoted B/A (pronounced 'B over A') and it is an abelian group under the operation

$$\bar{b}_1 + \bar{b}_2 = \overline{b_1 + b_2}.$$

This group is called the *quotient group* of B by A.

Apart from checking the group axioms it is necessary to check that the binary operation is well-defined, i.e. that if $\bar{b}_1 = \bar{b}_3$ and $\bar{b}_2 = \bar{b}_4$ then $\overline{b_1 + b_3} = \overline{b_2 + b_4}$. This is easily done using the following fact: $\bar{b} = \bar{b}'$ if and only if $b - b' \in A$.

Examples: (1) Let $B = \mathbb{Z}$ and $A = \{nk : n \in \mathbb{Z}\}$ where k is a fixed integer $\geqslant 1$. Thus A is the set of multiples of k. Then $f : \mathbb{Z}_k \to B/A$, defined by $fn = \bar{n}$ for $n = 0, 1, \ldots, k - 1$, is an isomorphism.

(2) $B/A = 0$ if and only if $B = A$. (In particular $0/0 = 0$.)

(3) If B is finite then $|B/A| = |B|/|A|$ since every coset has $|A|$ elements, the cosets are disjoint and there are $|B/A|$ of them. In particular $|A|$ divides $|B|$: this is Lagrange's theorem in the abelian group case.

A.14. Natural projection The *natural projection* $\pi : B \to B/A$ (where A is a subgroup of B) is defined by $\pi b = \bar{b}$ for all $b \in B$. Clearly π is an epimorphism (surjective homomorphism) with kernel A.

A.15. The homomorphism theorem *Let $f : G \to H$ be a homomorphism of abelian groups. Then $G/\mathrm{Ker} f \cong \mathrm{Im} f$, an isomorphism h being defined by $\bar{x} \to fx$ for all $x \in G$.*

Proof Firstly, h is well defined since

$$\bar{x} = \bar{y} \Rightarrow x - y \in \mathrm{Ker} f \Rightarrow f(x - y) = 0 \Rightarrow fx = fy$$

and a homomorphism since

$$h(\bar{x} + \bar{y}) = h(\overline{x + y}) = f(x + y) = fx + fy = h\bar{x} + h\bar{y}.$$

Finally h is monic (injective) since $fx = 0 \Rightarrow x \in \mathrm{Ker} f \Rightarrow \bar{x} = 0$ and epic (surjective) since $\mathrm{Im} f = \{fx : x \in G\}$. □

Examples: (1) Take $f : \mathbb{Z} \to \mathbb{Z}_k$ to be defined by $fn =$ residue (remainder) of n modulo k. Then f is a homomorphism with kernel the multiples of k. Compare example (1) in A.13.

★(2) Let G be the group of sequences $(\lambda_1, \lambda_2, \lambda_3, \ldots)$ where the λ_i are integers, under the operation $(\lambda_1, \lambda_2, \ldots) + (\mu_1, \mu_2, \ldots) = (\lambda_1 + \mu_1, \lambda_2 + \mu_2, \ldots)$. (Compare $\mathbb{Z}^n$ in example (3) of A.10.) Then G is an abelian group, not f.g. (in fact uncountable). Define $f : G \to G$ by $f(\lambda_1, \lambda_2, \lambda_3, \ldots) = (\lambda_2, \lambda_3, \ldots)$. Then f is epic, with kernel K isomorphic to $\mathbb{Z}$. Hence by the homomorphism theorem $G/K \cong G$; a nontrivial subgroup can be factored out without altering G up to isomorphism. It can be shown using A.27 below that that if A is f.g. and $A/B \cong A$ then $B = 0$ (see A.34(5)).★

Exact sequences

A.16. Exactness The 'sequence'

$$\cdots \to A \xrightarrow{f} B \xrightarrow{g} C \to \cdots$$

of abelian groups and homomorphisms is called *exact at* B if $\mathrm{Im} f = \mathrm{Ker} g$. (Note that this implies $g \circ f = 0$.) A sequence is called *exact* if it is exact at each group which is not at either end of the sequence. Exactness at an end group has no meaning. A *short exact sequence* is an exact sequence of the form

$$0 \to A \xrightarrow{f} B \xrightarrow{g} C \to 0$$

where, of course, the first and last homomorphisms are trivial.

The word 'exact' was first used in this sense in the book of Eilenberg & Steenrod listed in the References. There is some connexion with the concept of exact differential.

Examples: (1) $0 \to A \to 0$ is exact (i.e. exact at A) if and only if $A = 0$.

(2) $0 \to A \xrightarrow{f} B$ is exact if and only if f is monic (injective).

(3) $A \xrightarrow{f} B \to 0$ is exact if and only if f is epic (surjective).

(4) $0 \to A \xrightarrow{f} B \to 0$ is exact (at A and at B) if and only if f is an isomorphism.

(5) For any homomorphism $f : A \to B$, there is a short exact sequence

$$0 \to \operatorname{Ker} f \xrightarrow{i} A \xrightarrow{f'} \operatorname{Im} f \to 0.$$

Here i is the inclusion ($ix = x$ for all $x \in \operatorname{Ker} f$), so that f' is the same as f except for the target, and $f'a = fa$ for all $a \in A$.

(6) If A is a subgroup of B then

$$0 \to A \xrightarrow{i} B \xrightarrow{\pi} B/A \to 0$$

is exact, where $i =$ inclusion and $\pi =$ natural projection (A.14). Conversely if

$$0 \to A \xrightarrow{f} B \xrightarrow{g} C \to 0$$

is exact, then $C \cong B/\operatorname{Im} f$ by the homomorphism theorem A.15.

(7) If $A \subset B$ and B is a subgroup of C, then A is a subgroup of C and $B/A, C/A, C/B$ are defined. There is a short exact sequence

$$0 \to B/A \xrightarrow{i} C/A \xrightarrow{p} C/B \to 0$$

where $i(b+A) = b+A$ and $p(c+A) = c+B$. By the homomorphism theorem A.15, $C/B \cong (C/A)/(B/A)$: the cancellation rule for subgroups.

(8) If

$$0 \to A \to B \to C \to 0$$

is exact and B is finite, then A and C are also finite, and $|B| = |A| \cdot |C|$. (Compare A.13 Ex(3) and use $C \cong B/A$.)

Direct sums and splitting

A.17. Direct sum Let $A_1, \ldots, A_n$ be abelian groups. The *direct sum* $A_1 \oplus \cdots \oplus A_n$ of the groups is the set of n-tuples $(a_1, \ldots, a_n)$ where $a_1 \in A_1, \ldots a_n \in A_n$, under the binary operation

$$(a_1, \ldots, a_n) + (a'_1, \ldots, a'_n) = (a_1 + a'_1, \ldots, a_n + a'_n).$$

Then $A_1 \oplus \cdots \oplus A_n$ is an abelian group.

Direct sums and splitting 223

The most important case is $n = 2$, and there are 'associative' properties such as

$$A \oplus B \oplus C \cong (A \oplus B) \oplus C \cong A \oplus (B \oplus C)$$

which are very easy to check.

Examples: (1) The $\mathbb{Z}^n$ of A.10, Ex (3) is equal to $\mathbb{Z} \oplus \cdots \oplus \mathbb{Z}$ (n summands).

(2) If A and B are f.g. with generators $a_1, \ldots, a_n$ and $b_1, \ldots, b_m$ respectively, then $A \oplus B$ is f.g. with generators $(a_1, 0), \ldots, (a_n, 0), (0, b_1), \ldots, (0, b_m)$. If A and B are free and f.g. with the given generators as bases then $A \oplus B$ is free and f.g. with the given generators as basis. Similarly with more than two summands.

★(3) Suppose that p and q are coprime integers ≥ 1; consider $(1, 1) \in \mathbb{Z}_p \oplus \mathbb{Z}_q$. Now $n(1, 1) = (n, n)$ ($n \in \mathbb{Z}$) and $(n, n) = (0, 0) \Leftrightarrow p|n$ and $q|n \Leftrightarrow pq|n$ since p and q are coprime. Therefore $(1, 1)$ has order pq and $\mathbb{Z}_p \oplus \mathbb{Z}_q$, having order pq, is cyclic with $(1, 1)$ as generator. (Conversely if p and q have a common factor > 1, then $\mathbb{Z}_p \oplus \mathbb{Z}_q$ is not cyclic.) ★

A.18. Internal direct sum Let A and B be subgroups of C such that $A \cap B = \{0\}$. The *internal direct sum* $A \oplus B$ of A and B is the subgroup of C given by

$$A \oplus B = \{a + b : a \in A \text{ and } b \in B\}.$$

The clash of notation with A.17 is not serious since $(a, b) \to a + b$ defines an isomorphism between the direct sum and the internal direct sum of A and B, assuming of course $A \cap B = \{0\}$. Note that given c in the internal direct sum there exist *unique* $a \in A$ and $b \in B$ such that $c = a + b$.

Several theorems in the text are expressed by the exactness of certain sequences of homology groups. When applying these theorems, especially in Chapters 6, 7 and 9, the following result is often used.

A.19. Splitting lemma *Suppose that*

$$0 \to A \xrightarrow{f} B \xrightarrow{g} C \to 0$$

is exact.

(1) *If there exists a homomorphism* $h : C \to B$ *such that* $g \circ h$ *is the identity on C, then B is the internal direct sum (A.18) of* $\operatorname{Im} f$ *and* $\operatorname{Im} h$, *and furthermore* $B \cong A \oplus C$. *We call h a* splitting homomorphism *and say that the short exact sequence is* split.

(2) *If C is free and f.g., a splitting homomorphism always exists: the sequence is always split.*

Proof (1) Suppose that such an h exists. Then, for any $b \in B$, $(g \circ h \circ g)b = gb$, so that $b - (h \circ g)b \in \text{Ker}\, g = \text{Im}\, f$. Hence b can be written as the sum of elements from $\text{Im}\, f$ and $\text{Im}\, h$. To see that $\text{Im}\, f \cap \text{Im}\, h = \{0\}$ suppose that $fa = b = hc$ ($a \in A, b \in B, c \in C$). Then

$$0 = (g \circ f)a = gb = (g \circ h)c = c$$

by exactness and the given property of h. This gives the result, $b = 0$. Since f is monic, $\text{Im}\, f \cong A$ (for example using the homomorphism theorem A.15); likewise h is monic since $g \circ h = $ identity, so that $\text{Im}\, h \cong C$. Hence $B = \text{Im}\, f \oplus \text{Im}\, h \cong A \oplus C$.

(2) Let $c_1, \ldots, c_n$ be a basis for C and let $b_i \in B$ be chosen such that $gb_i = c_i, i = 1, \ldots, n$ (g is epic). Let $c \in C$. Then there exist unique integers $\lambda_1, \ldots, \lambda_n$ such that $c = \lambda_1 c_1 + \cdots + \lambda_n c_n$; define $hc = \lambda_1 b_1 + \cdots + \lambda_n b_n \in B$. Then h is a homomorphism and $g \circ h = $ identity. □

We shall make use at the end of this appendix of the following variant of the Splitting Lemma. Suppose for a moment that A, B, C are vector spaces over the same field and that f, g are linear maps. It still makes sense to say that

$$0 \to A \xrightarrow{f} B \xrightarrow{g} C \to 0$$

is exact: it means that f is injective, $\text{Im}\, f = \text{Ker}\, g$ (both are subspaces of B) and g is surjective. Suppose that C is finite-dimensional and let $c_1, \ldots, c_n$ be a basis for C.

As in (2) *above we can construct a linear map* $h : C \to B$ *such that* $g \circ h = $ *identity, and as in* (1) *we can show that this implies* $B = \text{Im}\, f \oplus \text{Im}\, h \cong A \oplus C$, *where this is the usual direct sum of vector spaces.* (In fact the same holds even if C is infinite-dimensional but the finite-dimensional case will suffice for the applications here.)

A.20. Examples (1) The sequence

$$0 \to A \xrightarrow{f} A \oplus C \xrightarrow{g} C \to 0$$

where $fa = (a, 0)$ and $g(a, c) = c$, is split by $h : C \to A \oplus C$ defined by $hc = (0, c)$. This applies whether C is free or not.

(2) Suppose that

$$0 \to \mathbb{Z} \xrightarrow{f} B \xrightarrow{g} \mathbb{Z}_2 \to 0$$

is exact. Let $a = f1$ and let $b \in B$ satisfy $gb = 1 \in \mathbb{Z}_2$. Then $g(2b) = 1+1 = 0$, so that $2b \in \operatorname{Ker} g = \operatorname{Im} f$ and $2b = \lambda a$ for some integer λ.
 Case I: λ is even. Say $\lambda = 2n$; then $h : \mathbb{Z}_2 \to B$ given by $h0 = 0, h1 = b - na$ is a splitting homomorphism, hence $B \cong \mathbb{Z} \oplus \mathbb{Z}_2$.
 Case II: λ is odd. Say $\lambda = 2n+1$; then $a = 2(b-na)$ and $b = (2n+1)(b-na)$. Hence B is cyclic, generated by $b - na$ (consider separately elements of B which go to 0 and 1 under g). But a has infinite order, so B is infinite: $B \cong \mathbb{Z}$ and f takes 1 to twice a generator of B. Note that any attempt to split the sequence will fail, for a splitting homomorphism h would have to satisfy $h1 + h1 = h(1+1) = 0$, and there is no element in $\mathbb{Z}$, besides 0, of finite order.

(3) Suppose that

$$0 \to A \xrightarrow{f} B \xrightarrow{g} C \to 0$$

is exact and that A and C are f.g. with generators $a_1, \ldots, a_n$ and $c_1, \ldots, c_m$. Let $b_i \in B$ satisfy $gb_i = c_i, i = 1, \ldots, m$. Then $fa_1, \ldots, fa_n, b_1, \ldots, b_m$ generate B, which is accordingly f.g.
 (Proof: let $b \in B$; then $gb = \lambda_1 c_1 + \cdots + \lambda_m c_m$ for suitable integers $\lambda_1, \ldots, \lambda_m$. Hence $b - \lambda_1 b_1 - \cdots - \lambda_m b_m \in \operatorname{Ker} g = \operatorname{Im} f$. Since $\operatorname{Im} f$ is generated by $fa_1, \ldots, fa_n$ this completes the proof.) Thus if the sequence is split, and A and C are known, then B is determined by the splitting lemma and generators for B are given by the above result. A more precise formulation when C is free is given in (4) below.
 If A and C are both free and f.g. and the above generators are bases then B is free with basis $fa_1, \ldots, fa_n, b_1, \ldots, b_m$.

(4) In the notation of the first part of (3) suppose that C is free, so that the sequences splits, and that $c_1, \ldots, c_m$ is a basis for C. Then I claim that there is an isomorphism $\theta : B \to A \oplus C$ for which $\theta(fa_i) = (a_i, 0), i = 1, \ldots, n; \theta b_i = (0, c_i), i = 1, \ldots, m$. (Proof below.) Thus $fa_1, \ldots, fa_n$ generate the subgroup of B corresponding under θ to $A \oplus 0$ while $b_1, \ldots, b_m$ generate the subgroup of B corresponding under θ to $0 \oplus C$. We often sum this up by:

$B \cong A \oplus C$ with generators $fa_1, \ldots, fa_n$ and $b_1, \ldots, b_m$.

Other examples of this usage occur in A.22 and A.25. (Proof of the claim: Let $b \in B$. Then there exist integers λ_i, μ_j such that $b = \sum \lambda_i fa_i + \sum \mu_j b_j$

by (3). Write $\theta b = (\sum \lambda_i a_i, \sum \mu_j c_j)$. It has to be checked that this is well-defined, and is an isomorphism. The first amounts to checking that if $b = 0$ then $\theta b = (0, 0)$. Now if $b = 0$ then $gb = 0$; since $g \circ f = 0$ this implies $\sum \mu_j c_j = 0$ and so $\mu_1 = \cdots = \mu_m = 0$ by the basis property. Since f is monic this gives $\sum \lambda_i a_i = 0$. Hence θ is well-defined. The rest of the proof is left to the reader.) The result may be false if C is not free, even if the sequence splits. E.g. in (1) above take $A = \mathbb{Z}, C = \mathbb{Z}_2$, and put $a_1 = 1$, $c_1 = 1$, $b_1 = (1, 1)$.

Presentations

Let A be a f.g. abelian group with generators $a_1, \ldots, a_n$. Let F be a free abelian group of rank n with basis $x_1, \ldots, x_n$ and let $t : F \to A$ be the epimorphism defined by $t(\lambda_1 x_1 + \cdots + \lambda_n x_n) = \lambda_1 a_1 + \cdots + \lambda_n a_n$ for all integers $\lambda_1, \ldots, \lambda_n$. (Note that this t is well-defined, since any element of F is uniquely expressible as an integer linear combination of $x_1, \ldots, x_n$.) The short exact sequence

$$0 \to \operatorname{Ker} t \overset{i}{\to} F \overset{t}{\to} A \to 0$$

(i being the inclusion) is sometimes called a 'presentation' of A; however we shall use this word in a different (but closely related) sense below. Note that by the homomorphism theorem we have $A \cong F/\operatorname{Ker} t$. Note also that there is usually a wide choice available for $a_1, \ldots, a_n$ and in particular for n.

The homology groups studied in the text do in fact arise in precisely this way, as the quotient of a free abelian group by a subgroup. It is worth looking more closely at what this means. Each element of the free abelian group F with basis $x_1, \ldots, x_n$ can be expressed uniquely as an integer linear combination of $x_1 \ldots, x_n$. Given a subgroup K of F, it follows from a later result (A.33) that K is free and f.g.; however in practice we are just given generators for K, so let $y_1, \ldots, y_m$ be any generators, not necessarily free, for K. Any element of F/K can be expressed as $\lambda_1 \bar{x}_1 + \cdots + \lambda_n \bar{x}_n$ for integers $\lambda_1, \ldots, \lambda_n$, but the expression is not unique (unless $K = 0$): in fact

$$\sum \lambda_i \bar{x}_i = \sum \mu_i \bar{x}_i \Leftrightarrow \sum (\lambda_i - \mu_i)\bar{x}_i = 0 \Leftrightarrow \sum (\lambda_i - \mu_i) x_i \in K.$$

Now $x_i, \ldots, x_n$ is a basis for F so there are uniquely determined integers α_{ij} ($i = 1, \ldots, m; j = 1, \ldots, n$) such that

$$y_i = \alpha_{i1} x_1 + \cdots + \alpha_{in} x_n \quad (i = 1, \ldots, m)$$

and an element of F belongs to K if and only if it is an integer linear combination of the y's. The group F/K is then completely specified by the information that

$$\bar{x}_1,\ldots,\bar{x}_n \quad \text{generate it}$$

and that

$$\alpha_{i1}\bar{x}_1 + \cdots + \alpha_{in}\bar{x}_n = 0 \text{ for } i = 1,\ldots,m;$$

for the equations tell us precisely when two integer linear combinations of the $\bar{x}$'s give the same element of F/K.

The equations are called *relations* and the above description of F/K is called a presentation (by generators and relations). In practice we drop the bars, which are a notational inconvenience, and call

A.21
$$\begin{pmatrix} x_1,\ldots,x_n \\ \alpha_{i1}x_1 + \cdots + \alpha_{in}x_n = 0 \quad (i = 1,\ldots,m) \end{pmatrix}$$

a *presentation* of F/K.

The formulation just given is designed for situations where, as with the definition of homology groups, we are given a free abelian group F and a subgroup K and wish to identify F/K. It is also possible to start with a f.g. abelian group A, choose F and t as at the beginning of the section, and let $K = \text{Ker } t$. Then replacing each x_i by $tx_i = a_i \in A$ in A.21 we obtain what can be called a presentation of A. For example let $A = \mathbb{Z} \oplus \mathbb{Z}_2$, $a_1 = (1,0)$, $a_2 = (0,1)$. Then

$$\begin{pmatrix} a_1, & a_2 \\ & 2a_2 = 0 \end{pmatrix}$$

is in this sense a presentation of $\mathbb{Z} \oplus \mathbb{Z}_2$. (Compare A.28(1).) We shall stick to the former interpretation in what follows.

A.22. Example Consider the abelian group F/K with presentation

$$\begin{pmatrix} x,y \\ \alpha x + \beta y = 0 \end{pmatrix}$$

where α and β are integers, not both zero. Let h be the greatest common divisor of α and β (the greatest common divisor of 0 and an integer k is, naturally, defined to be $|k|$), and let $\alpha = \alpha_1 h, \beta = \beta_1 h$ so that α_1 and β_1 are coprime. Hence there exist integers γ, δ such that $\alpha_1\gamma + \beta_1\delta = 1$; write

$$\alpha_1 x + \beta_1 y = u \quad \text{and} \quad -\delta x + \gamma y = v.$$

Then $x = \gamma u - \beta_1 v$; $y = \delta u + \alpha_1 v$ so that u and v generate the free abelian group F with basis x, y and so form a basis for it. (This can be checked directly;

the general case follows from A.30 and A.34(2).) In terms of the basis u, v the relation becomes $hu = 0$, so that

$$\begin{pmatrix} u, & v \\ hu = 0 & \end{pmatrix}$$

is a presentation of the same abelian group F/K. Now this presentation is easy to interpret, for the epimorphism

$$F \to \mathbb{Z}_h \oplus \mathbb{Z}$$

given by $\quad \lambda u + \mu v \to (\lambda, \mu) \quad (\lambda, \mu \in \mathbb{Z})$

has kernel the subgroup K generated by hu. Hence

$$F/K \cong \mathbb{Z}_h \oplus \mathbb{Z}.$$

Furthermore the elements of F/K which correspond to generators $(1, 0)$ and $(0, 1)$ of $\mathbb{Z}_h \oplus \mathbb{Z}$ are $\bar{u}$ and $\bar{v}$. We express this by saying $F/K \cong \mathbb{Z}_h \oplus \mathbb{Z}$ with generators $\bar{u}$ and $\bar{v}$. (Compare A.20(4).)

When we are given a presentation of an abelian group as in the above example, the problem is to write the group in some recognizable form. This involves changing the generators and relations, without changing the group which they present, until they become simple enough to spot the group up to isomorphism. Here are three ways in which generators and relations presenting a group F/K can be changed.

A.23. New generators and relations for old (1) We can take a new basis for F and re-express the relations in terms of the new basis. (If F has a basis containing n elements, then any n elements of F which generate F automatically form a basis – see A.30, A.34(2).) This was done in A.22, where x and y became u and v. Neither F nor K is changed, so F/K certainly remains the same.

(2) We can change the generators for K. This amounts to replacing the relations by an equivalent set – that is, one which implies and is implied by the old set. For instance we could make a succession of changes of the following kind.

$$i\text{th relation} \to \pm i\text{th relation} + \lambda(j\text{th relation})$$

where λ is an integer and $i \neq j$. Certainly the new relations will be equivalent to the old ones and the 'left-hand sides' will define the same K. Note that the relation $0 = 0$ can be removed if ever it occurs.

(3) Consider the presentations

$$\begin{pmatrix} x_1, \ldots, x_k \\ r_1 = 0, \ldots, r_m = 0 \end{pmatrix} \text{ and } \begin{pmatrix} x_1, \ldots, x_k, & x_{k+1}, \ldots, x_n \\ r_1 = 0, \ldots, r_m = 0 \\ x_{k+1} = 0, \ldots, x_n = 0 \end{pmatrix}$$

(Here the r's are of course integer linear combinations of $x_1, \ldots, x_k$.) All that has happened here is that some extra generators have been added, and put equal to zero – given with one hand and taken away with the other. Writing the groups presented as F_1/K_1 and F/K respectively, we can regard F_1 as a subgroup of F; then K_1 is a subgroup of K and the inclusion $F_1 \subset F$ gives an isomorphism $F_1/K_1 \cong F/K$.

A.24. Further changes of presentation The next change of presentation involves elements from all the above. Suppose we have a presentation of F/K with generators.

$$x_1, \ldots, x_k, \quad x_{k+1}, \ldots, x_n$$

and relations

(R_1)
$$\begin{aligned} x_{k+1} - \text{(integer linear combination of } x_1, \ldots, x_k\text{)} &= 0 \\ &\vdots \\ x_n - \text{(integer linear combination of } x_1, \ldots, x_k\text{)} &= 0 \end{aligned}$$

(R_2) Other relations, possibly involving all of $x_1, \ldots, x_n$.

Then (R_1) tells us that $x_{k+1}, \ldots, x_n$ are redundant generators; we proceed to eliminate them. The relations (R_2) can be expressed entirely in terms of $x_1, \ldots, x_k$ by substituting from (R_1) for $x_{k+1}, \ldots, x_n$. Let the result be (R'_2). By A.23(2) the presentation with generators $x_1, \ldots, x_n$ and relations (R_1) and (R'_2) defines the same F/K. Now let $x'_{k+1}, \ldots, x'_n$ denote the left-hand sides of the relations (R_1). Then $x_1, \ldots, x_k, x'_{k+1}, \ldots, x'_n$ are also a basis for F so that using (1) the following presentation gives the same F/K.

$$\begin{pmatrix} x_1, \ldots, x_k, & x'_{k+1}, \ldots, x'_n \\ (R'_1) \; x'_{k+1} = 0, \ldots, x'_n = 0 \\ (R'_2) \text{ As } (R_2), \text{ but with } x_{k+1}, \ldots, x_n \text{ eliminated by using } (R_1). \end{pmatrix}$$

Finally we invoke (3) to say that F_1/K_1 defined by

$$\begin{pmatrix} x_1, \ldots, x_k \\ (R'_2) \end{pmatrix}$$

is isomorphic to F/K, an isomorphism being induced by the inclusion $F_1 \subset F$.

Here is an example which is typical of the calculations in the text.

A.25. Example Consider F/K defined by twelve generators $x_1, \ldots, x_{12}$ and eleven relations as follows:

$$\begin{aligned} x_1 - x_8 &= 0 & x_4 + x_5 &= 0 \\ x_2 - x_8 + x_9 &= 0 & x_5 + x_6 &= 0 \\ x_9 - x_{10} - x_{11} &= 0 & x_6 + x_7 - x_{12} &= 0 \\ x_{11} - x_{12} &= 0 & x_1 - x_7 &= 0 \\ x_2 + x_3 &= 0 & x_{10} &= 0 \\ x_3 + x_4 &= 0 \end{aligned}$$

First write the relations in the following equivalent form.

$$\begin{aligned} x_8 &= x_1 & x_6 &= x_2 \\ x_7 &= x_1 & x_{10} &= 0 \\ x_3 &= -x_2 & x_{11} &= x_9 \\ x_4 &= x_2 & \text{(from } x_{11} &= x_9 - x_{10} = x_9) \\ \text{(from } x_4 &= -x_3 = x_2) & x_{12} &= x_9 \\ x_5 &= -x_2 & \text{(from } x_{12} &= x_{11} = x_9) \\ & & x_6 + x_7 - x_{12} &= 0 \\ & & x_2 - x_8 + x_9 &= 0 \end{aligned}$$

This makes it clear that all generators except x_1, x_2 and x_9 are redundant, and invoking A.24 (the first nine relations are (R_1) and the last two are (R_2)) we are reduced to

$$\begin{pmatrix} x_1, x_2, x_9 \\ x_2 + x_1 - x_9 = 0 \\ x_2 - x_1 + x_9 = 0 \end{pmatrix}$$

Eliminating x_9 in the same way we obtain

$$\begin{pmatrix} x_1, x_2 \\ x_2 - x_1 + (x_2 + x_1) = 0 \end{pmatrix}, \quad \text{i.e.} \quad \begin{pmatrix} x_1, x_2 \\ 2x_2 = 0 \end{pmatrix}$$

The group F_1/K_1 now presented is clearly $\mathbb{Z} \oplus \mathbb{Z}_2$; there is an exact sequence

$$0 \to K_1 \xrightarrow{i} F_1 \xrightarrow{p} \mathbb{Z} \oplus \mathbb{Z}_2 \to 0$$

where i is inclusion and $p(x_1) = (1,0), p(x_2) = (0,1)$. Since $F_1 \subset F$ induces $F_1/K_1 \cong F/K$ we deduce that $F/K \cong \mathbb{Z} \oplus \mathbb{Z}_2$ and that an isomorphism $F/K \to \mathbb{Z} \oplus \mathbb{Z}_2$ is given by $\bar{x}_1 \to (1,0)$, $\bar{x}_2 \to (0,1)$, bars denoting cosets with respect to K. We express this by saying '$F/K \cong \mathbb{Z} \oplus \mathbb{Z}_2$ with generators $\bar{x}_1$ and $\bar{x}_2$'. (Compare A.20(4).)

What has been said about presentations so far should enable the reader to follow the comparatively few direct calculations of homology groups from presentations in the text, e.g. 4.15, 4.29(1) and (3). It is possible to give a procedure for reducing to a 'standard form' any given presentation; this procedure is based on A.22–24 and is usually described in terms of row and column operations on the 'relation matrix' (α_{ij}) of A.21. For the kind of calculations required in this book the general method is unnecessarily cumbersome, and it is easier to operate directly on the relations as in A.25. However the general method enables one to prove general theorems. The following theorem and corollary, which are given here without proof, receive full discussion in textbooks of algebra, for example, the books by Lang and Mac Lane & Birkhoff listed in the References.

A.26. Theorem *Let F be a free f.g. abelian group and K a subgroup of F. Then F/K has a presentation of the form*

$$\begin{pmatrix} x_1, \ldots, x_n \\ \delta_1 x_1 = 0, \ldots, \delta_k x_k = 0 \end{pmatrix}$$

where $\delta_1, \ldots, \delta_k$ are integers > 1 $\delta_1|\delta_2, \delta_2|\delta_3, \ldots, \delta_{k-1}|\delta_k$. (Of course $k \leq n$. If there are no relations we say $k = 0$.) The integer $n - k$, and the δ_i, are determined by the isomorphisms class of F/K. □

We call $n - k$ the *rank* of F/K and the δ_i are called the *invariant factors* or *torsion coefficients* of F/K. An alternative approach to rank is described in the next section. The above presentation is sometimes called the *Smith normal form*.

A.27. Corollary: Classification theorem for f.g. abelian groups

$$F/K \cong \mathbb{Z}_{\delta_1} \oplus, \cdots \oplus \mathbb{Z}_{\delta_k} \oplus \mathbb{Z} \oplus \cdots \oplus \mathbb{Z}$$

where there are $n - k$ $\mathbb{Z}$'s, i.e. rank $F/K = n - k$. □

Note that from the discussion leading to A.21 every f.g. abelian group is isomorphic to such a quotient F/K. The corollary follows from the theorem by examining the short exact sequence

$$0 \to K \to F \xrightarrow{t} \mathbb{Z}_{\delta_1} \oplus \cdots \oplus \mathbb{Z}_{\delta_k} \oplus \mathbb{Z} \oplus \cdots \oplus \mathbb{Z} \to 0$$

where $tx_i = (0, \ldots, 0, 1, 0, \ldots, 0)$ with the 1 in the ith position, for $i = 1, \ldots, n$.

A.28. Remarks and examples (1) It is also possible to use the method of presentations to calculate quotients B/A where B is not free. In fact if B is merely f.g. then, roughly speaking, we take a presentation for B and add extra relations to express the fact that generators for A are zero. For example take $B = \mathbb{Z} \oplus \mathbb{Z}_2$ and $A =$ subgroup generated by $(3, 1)$. Then writing $a = (1, 0), b = (0, 1)$ we have the presentation

$$\left(\begin{array}{c} a, b \\ 2b = 0 \\ 3a + b = 0 \end{array} \right.$$

for B/A. This gives $B/A \cong \mathbb{Z}_6$ with generator $a + A$. (Compare the example following A.21.)

Calculations of the kind just given can be an aid when using the homomorphism theorem A.15, for if $g : B \to C$ is a homomorphism then $\text{Im } g \cong B/\text{Ker } g$. An example occurs towards the end of 7.5, though it so happens that there B is free.

★(2) The classification theorem A.27 implies at once that a f.g. abelian group containing no element of finite order is free. This is not easy to prove directly – indeed it is a good part of the content of A.27. For given any abelian group A the subset T of elements of finite order forms a subgroup, called the *torsion subgroup*, of A. There is a short exact sequence

$$0 \to T \to A \to A/T \to 0.$$

Certainly A/T has no element of finite order so, assuming that A is f.g., the above result implies that A/T is free and the sequence splits; $A \cong T \oplus A/T$. This is the decomposition of A.27 without the detailed structure of T. ★

Rank of a f.g. abelian group

Some of the results below can be obtained from the classification theorem A.27, but since we do not prove that theorem a self-contained discussion is given here. Except for starred examples, what follows assumes only A.1–A.20.

Let A be (as usual) an abelian group. We shall construct a vector space $\tilde{A}$ over the rational numbers $\mathbb{Q}$ which, roughly speaking, suppresses the elements of finite order in A. Some results needed in the text do not involve the elements of finite order, and these results can be proved readily by this means. The construction is a special case of the 'tensor product' construction (see for example the books of Lang and Mac Lane & Birkhoff listed in the references) but in the present case the details are much more straightforward than in the general case, and in fact follow precisely the standard construction of the rational numbers from the integers. In tensor notation, we are constructing $A \otimes \mathbb{Q}$.

The elements of $\tilde{A}$ are symbols ('fractions') a/n, pronounced 'a over n', where $a \in A$ and n is a non-zero integer. (The tensor notation is $a \otimes \frac{1}{n}$. Beware that in a general tensor product $A \otimes B$ not every element can be written in the form $a \otimes b$ with $a \in A$ and $b \in B$.) We identify a/n with b/m if and only if, for some non-zero integers k, ℓ, we have

$$ka = \ell b, \quad kn = \ell m.$$

(This can be expressed alternatively by saying that we impose the corresponding equivalence relation on pairs $(a, \frac{1}{n})$ and then work with the equivalence classes.) In particular $a/n = ka/kn$ for any non-zero integer k.

The rules for addition and scalar multiplication in $\tilde{A}$ are:

$$a_1/n_1 + a_2/n_2 = (n_2 a_1 + n_1 a_2)/n_1 n_2$$

$r(a/n) = pa/qn$ where $r = p/q$ is a rational

number and p, q are integers.

These rules respect the identifications, and $\tilde{A}$ is a rational vector space. The zero vector is $0/1$. Note the following important fact: $a/n = 0/1$ if and only if a has finite order. (Proof: if a has finite order $k \geqslant 1$, so that $ka = 0$, then $a/n = ka/kn = 0/kn = 0/1$. If $a/n = 0/1$ then, for some non-zero integers k, ℓ we have $ka = 0$, $kn = \ell$. Hence a has finite order.) Thus passing from A to $\tilde{A}$ suppresses the elements of finite order.

A.29. Rank Let A be f.g. with generators $a_1, \ldots, a_n$. Then $a_1/1, \ldots, a_n/1$ generate $\tilde{A}$ as a rational vector space, and accordingly $\tilde{A}$ is finite dimensional.

The *rank* of A is defined to be the dimension of $\widetilde{A}$. (This is reconciled with A.26 and A.27 in A.32(2) below.)

A.30. Theorem *Let A be free and f.g. with basis $a_1, \ldots, a_n$. Then rank $A = n$ and every basis for A has n elements.*

Proof We show that $a_1/1, \ldots, a_n/1$ is a basis for $\widetilde{A}$, which accordingly has dimension n so that rank $A = n$. The second result follows because every basis for $\widetilde{A}$ contains n elements.

Certainly, as in A.29, the stated elements generate $\widetilde{A}$. For $a \in A$ has the form $\lambda_1 a_1 + \cdots + \lambda_n a_n$ for integers λ_i, and we have $a/m = \lambda_1/m \cdot a_1/1 + \cdots + \lambda_n/m \cdot a_n/1$ so that every element of $\widetilde{A}$ is a rational linear combination of $a_1/1, \ldots, a_n/1$. To see that these vectors are linearly independent, suppose that, for some rational numbers $r_1, \ldots, r_n$, we have $r_1(a_1/1) + \cdots + r_n(a_n/1) = 0/1$. Multiplying through by the product, k say, of the denominators of the r_i we obtain $kr_1(a_1/1) + \cdots + kr_n(a_n/1) = 0/1$, where the coefficients are now integers. Hence $kr_1 a_1 + \cdots + kr_n a_n$ has finite order and, A being free, must in fact be 0. Using $k \neq 0$ and the basis property we have $r_1 = \cdots = r_n = 0$. □

Examples: $\mathbb{Z}^n$ has rank n; the trivial group has rank 0; $\mathbb{Z} \oplus \mathbb{Z}_2$ has rank 1.

Now let $f : A \to B$ be a homomorphism. Define $\widetilde{f} : \widetilde{A} \to \widetilde{B}$ by $\widetilde{f}(a/n) = fa/n$. This is well-defined (i.e. if $a/n = b/m$ then $fa/n = fb/m$) and is a linear map from $\widetilde{A}$ to $\widetilde{B}$. (What we have now defined is a 'functor' from the 'category' of abelian groups and homomorphisms to the 'category' of rational vector spaces and linear maps. Information on these concepts is in many books of abstract algebra, for example the book of Mac Lane & Birkhoff listed in the References. The next remark says that the functor is 'exact'.)

Suppose that

$$A \xrightarrow{f} B \xrightarrow{g} C$$

is exact (at B). Then

$$\widetilde{A} \xrightarrow{\widetilde{f}} \widetilde{B} \xrightarrow{\widetilde{g}} \widetilde{C}$$

is exact (at $\widetilde{B}$), i.e. the subspaces $\operatorname{Im}\widetilde{f}$ and $\operatorname{Ker}\widetilde{g}$ of $\widetilde{B}$ are equal.

(Proof: $\widetilde{g}\widetilde{f}(a/n) = \widetilde{g}(fa/n) = gfa/n = 0/n = 0/1$ so $\operatorname{Im}\widetilde{f} \subset \operatorname{Ker}\widetilde{g}$. Suppose $\widetilde{g}(b/n) = 0/1$. Then gb has finite order k say, and $g(kb) = k(gb) = 0$. Hence $kb \in \operatorname{Ker} g = \operatorname{Im} f$, i.e. $kb = fa$ for some $a \in A$. Then $b/n = kb/kn = fa/kn = \widetilde{f}(a/kn)$. Hence $\operatorname{Ker}\widetilde{g} \subset \operatorname{Im}\widetilde{f}$.) Thus the operation $\sim$ takes exact sequences to exact sequences. Recall from the discussion before A.20 that every short exact sequence of vector spaces and linear maps is split.

A.31. Theorem *Suppose that*

$$0 \to A \xrightarrow{f} B \xrightarrow{g} C \to 0$$

is an exact sequence of f.g. abelian groups. Then

$$\operatorname{rank} B = \operatorname{rank} A + \operatorname{rank} C.$$

Proof The corresponding short exact sequence of vector spaces splits, so $\widetilde{B} \cong \widetilde{A} \oplus \widetilde{C}$; hence

$$\operatorname{rank} B = \dim \widetilde{B} = \dim(\widetilde{A} \oplus \widetilde{C}) = \dim \widetilde{A} + \dim \widetilde{C} = \operatorname{rank} A + \operatorname{rank} C.$$

$\square$

A.32. Examples (1) If A and C are f.g. then $\operatorname{rank}(A \oplus C) = \operatorname{rank} A + \operatorname{rank} C$. (Apply A.31 to the short exact sequence of A.20(1).)

(2) We are now in a position to see that, defining $\operatorname{rank} A = \dim \widetilde{A}$ as in A.29, the rank of the group

$$A = \mathbb{Z}_{\delta_1} \oplus \cdots \oplus \mathbb{Z}_{\delta_k} \oplus \mathbb{Z}^{n-k}$$

of A.27 really is $n - k$. For $A = T \oplus \mathbb{Z}^{n-k}$ where T is finite (i.e. $\widetilde{T} = 0$) and we have

$$\operatorname{rank} A = \operatorname{rank} T + \operatorname{rank} \mathbb{Z}^{n-k} \text{ by (1) above}$$
$$= 0 + (n - k) \text{ using } \widetilde{T} = 0 \text{ and A.30}.$$

(3) Let

$$0 \to A_1 \xrightarrow{f_1} A_2 \xrightarrow{f_2} \cdots \to A_{n-1} \xrightarrow{f_{n-1}} A_n \xrightarrow{f_n} 0$$

be an exact sequence of f.g. abelian groups. We can define $n - 1$ short exact sequences

$$0 \to \operatorname{Ker} f_i \to A_i \to \operatorname{Im} f_i \to 0 \quad (i = 1, \ldots, n-1)$$
$$\phantom{0 \to \operatorname{Ker} f_i \to A_i \to} \| $$
$$\phantom{0 \to \operatorname{Ker} f_i \to A_i \to} \operatorname{Ker} f_{i+1}$$

(Compare A.16(5).) Applying A.31 to each of these yields the *alternating sum of ranks theorem*: $\sum_{i=1}^{n} (-1)^i \operatorname{rank} A_i = 0$.

A.33. Theorem *Suppose that B is free and f.g. of rank m, and that A is a subgroup of B. Then A is free and f.g. of rank $\leqslant m$.*

Proof We proceed by induction on m. The result for $m = 1$ follows from the fact that any nontrivial subgroup of $\mathbb{Z}$ consists of the multiples of a fixed non-zero integer and is therefore itself infinite cyclic. Now assume the result is proven when B has rank $< m$ ($m \geqslant 2$) and chose a basis $b_1, \ldots, b_m$ for the free group B of rank m. Define $t : B \to \mathbb{Z}$ by $t(\lambda_1 b_1 + \cdots + \lambda_m b_m) = \lambda_m$; then $\operatorname{Ker} t$ is free of rank $m - 1$, with basis $b_1, \ldots, b_{m-1}$. We shall apply the induction hypothesis to $\operatorname{Ker} t$ by letting $C = tA$ and $t' : A \to C$ be the epimorphism given by $t'a = ta$ for all $a \in A$. Then $\operatorname{Ker} t'$ is a subgroup of $\operatorname{Ker} t$ and is therefore free of rank $\leqslant m - 1$ by the induction hypothesis. The short exact sequence

$$0 \to \operatorname{Ker} t' \xrightarrow{i} A \xrightarrow{t'} C \to 0$$

is split (A.19) since C is a subgroup of $\mathbb{Z}$ and therefore free of rank $\leqslant 1$. Thus $A \cong \operatorname{Ker} t' \oplus C$ is free of rank $\leqslant m$. (Once A is known to be f.g. the inequality of ranks follows also from the fact that $\widetilde{A}$ is a subspace of $\widetilde{B}$.) □

A.34. Examples (1) Suppose that A and B are free and f.g. of the same rank and that $f : A \to B$ is an epimorphism (surjective). Then f is in fact an isomorphism. (Consider the short exact sequence $0 \to \operatorname{Ker} f \xrightarrow{i} A \xrightarrow{f} B \to 0$. This splits by A.19 and therefore $A \cong \operatorname{Ker} f \oplus B$. Hence $\operatorname{rank}(\operatorname{Ker} f) = 0$ and since $\operatorname{Ker} f$ is free by A.33 it must be zero, i.e. f is monic (injective).) In contrast $f : \mathbb{Z} \to \mathbb{Z}$ defined by $fn = 2n$ is monic without being epic.

(2) Suppose that A is free and f.g. of rank n and that $a_1, \ldots, a_n$ generate A. Then in fact they are a basis for A. (This is just (1) applied to the epimorphism $\mathbb{Z}^n \to A$ given by $(\lambda_1, \ldots, \lambda_n) \to \lambda_1 a_1 + \cdots + \lambda_n a_n$.)

(3) Let A be free and f.g. and suppose that a and b are elements of A with the property that $\alpha a = \beta b$ for some integers α, β not both zero. Then a and b have a 'common factor', i.e. there is an element $c \in A$ of which a and b are both multiples. (When $A = \mathbb{Z}$ the hypothesis $\alpha a = \beta b$ is redundant since we can take $\beta = a, \alpha = b$, unless $a = b = 0$ when we can take $\alpha = \beta = 1$.) To see this suppose $a \neq 0$, so that $\beta \neq 0$, and let C be the subgroup of A generated by a and b. Denoting by $\langle a \rangle$ the infinite cyclic subgroup of C generated by a, it follows that $C/\langle a \rangle$ is finite, since a non-zero multiple βb of b belongs to $\langle a \rangle$. Hence $\operatorname{rank} C = 1$ by A.31, i.e. C is infinite cyclic and a suitable c is a generator for C.

(4) A non-zero element a of an abelian group A is called *indivisible* if it is not a multiple of any element of A besides $\pm a$ (i.e. if $a = \lambda b$, $\lambda \in \mathbb{Z}$, $b \in A \Rightarrow \lambda = \pm 1$). Thus for example the additive group of rational numbers $\mathbb{Q}$ contains no indivisible element, and the only indivisible elements of $\mathbb{Z}$ are ± 1. Let A be a free f.g. abelian group, and let $a \in A$ be non-zero. Then *the following four statements are equivalent.*

(i) *a is indivisible.*

(ii) *$A/\langle a \rangle$ is free abelian, $\langle a \rangle$ being the (infinite cyclic) subgroup of A generated by a.*

(iii) *a is an element of some basis for A.*

(iv) *Given any basis $x_1, \ldots, x_n$ for A the unique expression $a = \lambda_1 x_1 + \cdots + \lambda_n x_n$ ($\lambda_1, \ldots, \lambda_n \in \mathbb{Z}$) has the greatest common divisor of $\lambda_1, \ldots, \lambda_n$ equal to 1.*

(The proofs (i) $\Rightarrow$ (ii) $\Rightarrow$ (iii) $\Rightarrow$ (iv) $\Rightarrow$ (i) are fairly straightforward. Here are some hints. (i) $\Rightarrow$ (ii): show $A/\langle a \rangle$ has no non-zero element of finite order, perhaps using (3) above. (ii) $\Rightarrow$ (iii): Use the short exact sequence $0 \to \langle a \rangle \to A \to A/\langle a \rangle \to 0$. (iii) $\Rightarrow$ (iv): Let $a_1 = a, a_2, \ldots, a_n$ be a basis containing a. Let Λ be the $n \times n$ integral matrix expressing the a's in terms of the x's; thus the first row of Λ is $\lambda_1, \ldots, \lambda_n$. Since the a's are a basis we can consider the matrix Λ^{-1} expressing the x's in terms of the a's, and we have $\Lambda \Lambda^{-1} = I_n$, the $n \times n$ identity matrix. Hence $\det \Lambda \cdot \det \Lambda^{-1} = 1$, so $\det \Lambda = \pm 1$ (i.e. Λ is unimodular). Expanding $\det \Lambda$ by the first row shows that the greatest common divisor of $\lambda_1, \ldots, \lambda_n$ is 1. (iv) $\Rightarrow$ (i): follows from the definition of basis.)

★(5) Let $f : A_1 \to A_2$ be a homomorphism of abelian groups and let T_i be the torsion subgroup of A_i ($i = 1, 2$), that is the subgroup of elements of finite order. Then we have $f(T_1) \subset T_2$ so f restricts to a homomorphism $f' : T_1 \to T_2$ defined by $f't = ft$ for all $t \in T_1$. (This is another example of a functor (compare the discussion following A.30), this time from the category of abelian groups and homomorphism to itself. This functor is not exact, as can be verified by considering the exact sequence $\mathbb{Z} \to \mathbb{Z} \oplus \mathbb{Z}_2 \to \mathbb{Z}$ where the homomorphisms are $m \to (0, m \bmod 2)$ and $(n, p) \to n$. But read on.)

Suppose that A_1 is f.g. and $\mathrm{Ker} f$ is finite. Then the exactness of

$$A_1 \xrightarrow{f} A_2 \xrightarrow{g} A_3$$

implies the exactness of

$$T_1 \xrightarrow{f'} T_2 \xrightarrow{g'} T_3.$$

Hints for proof. Clearly $g' \circ f' = 0$. Identify T_1 with $0 \oplus T_1$ and use A.27 to write $A_1 \cong F_1 \oplus T_1$ with F_1 free and f.g. For simplicity write $A_1 = F_1 \oplus T_1$. Now

$g't_2 = 0 \Rightarrow t_2 = f(x_1, t_1)$ for some $(x_1, t_1) \in A_1$. Hence $t_2 = f(x_1, 0) + f(0, t_1)$. Deduce $f(x_1, 0)$ has finite order, k say. Hence $(kx_1, 0) \in \text{Ker} f$ but has infinite order unless $x_1 = 0$.)

As an example suppose B is a subgroup of the f.g. abelian group A and suppose $A/B \cong A$. We show $B = 0$. Using the exact sequence

$$0 \to B \xrightarrow{i} A \xrightarrow{\pi} A/B \to 0$$

(compare A.16 Ex (6)), and A.31, it follows that rank $B = 0$, i.e. B is finite. The result above then shows that

$$0 \to B \to T \to T' \to 0$$

is exact where T, T' are the torsion subgroups of A and A/B respectively. But T and T' have the same number of elements, so $B = 0$. ★

★(6) Consider an exact sequence

$$A \xrightarrow{f} A \oplus B \xrightarrow{g} B.$$

(i) If A and B are finite, then f is monic and g is epic. (Construct two short exact sequences from f and g as in A.16 Ex(5) and use A.16 Ex(8).)

(ii) If A and B are f.g. then Ker f is finite. (Show this and $B/\text{Im } g$ is finite by applying the alternating sum of ranks theorem (A.32(3)) to the exact sequence $0 \to \text{Ker} f \to A \to A \oplus B \to B \to B/\text{Im } g \to 0$.) Hence if A, B are f.g. then with f' as in (5) above we have Ker $f = \text{Ker} f' = 0$ by (5) and (i). Hence: if A and B are f.g. then f is monic. It is not true in general that g is epic.

(iii) Take $B = 0$ in (ii). We have: if A is f.g. then any epimorphism $A \to A$ is an isomorphism. (Compare A.34(1). It is not hard to construct counterexamples when A is not f.g.)

(iv) If A is f.g. and B is finite then f is monic and g is epic. ★

References

Apart from books specifically cited in the text, the following list contains a few others which can be regarded as parallel, or further, reading (possibly both). In the latter respect the list is by no means exhaustive. Dates refer to recent editions (at the time of writing), but a small number of the older books listed may be available only second-hand or through libraries. Books which could be regarded as parallel reading are prefixed by †. More advanced books are prefixed by ‡. Articles in journals and edited books are listed separately, below.

Books

‡Ahlfors, L. V. & L. Sario, *Riemann Surfaces*, Princeton University Press (1960).
†Aldous, J. M., R. J. Wilson & S. Best, *Graphs and Applications*, Springer-Verlag (2003).
†Berge, C. *The Theory of Graphs*, Dover Publications (2003).
Biggs, N. L., E. K. Lloyd & E. S. Wilson, *Graph Theory 1736–1936*, Oxford University Press (1976, 1986).
‡Bollobás, B. *Modern Graph Theory*, Springer-Verlag (1998, 2002).
†Courant, R. & H. E. Robbins, revised by Ian Stewart *What is Mathematics?*, Oxford University Press (1996).
‡Diestel, R. *Graph Theory*, Springer-Verlag (2006).
Dieudonné, J. *A History of Algebraic and Differential Topology 1900–1960*, Birkhäuser (2009).
‡Eilenberg, S. & N. E. Steenrod, *Foundations of Algebraic Topology*, Princeton University Press (1952).
† Firby, P. A. & C. F. Gardiner, *Surface Topology*, Ellis Horwood (2001).
†Flegg, H. G. *From Geometry to Topology*, Dover Publications (2003).
†Gross, J. & J. Yellen, *Graph Theory and its Applications*, CRC Press (2005).
†Hilbert, D. & S. Cohn-Vossen, *Geometry and the Imagination*, Chelsea Publishing Company/American Mathematical Society (1999).
‡Hilton, P. J. & S. Wylie, *Homology Theory*, Cambridge University Press (1968).
‡Hocking, J. G. & G. S. Young, *Topology*, Dover Publications (1989).
‡Kelley, J. L. *General Topology*, Springer-Verlag (1975).
Kline, M. *Mathematical Thought from Ancient to Modern Times*, Oxford University Press (1990).

Lakatos, I., J. Worrall & E. Zahar, *Proofs and Refutations*, Cambridge University Press (1976).
‡Lang, S. *Algebra*, Springer-Verlag (2002).
‡Lawson, T. *Topology: A Geometric Approach*, Oxford Graduate Texts No. 9, Oxford University Press (2003, 2006).
†Mac Lane, S. & G. D. Birkhoff, *Algebra*, Chelsea Publishing Company/American Mathematical Society (1988).
†Massey, W. S. *Algebraic Topology: An Introduction*, Graduate Texts in Mathematics, vol. 56, Springer-Verlag (1989).
‡Maunder, C. R. F. *Algebraic Topology*, Dover Publications (1996).
†Milnor, J. W. *Topology from the Differentiable Viewpoint*, Princeton University Press (1997).
‡Milnor, J. W. & J. D. Stasheff, *Characteristic Classes*, Annals of Mathematics Studies No. 76, Princeton University Press (1974).
†Newman, M. H. A. *Elements of the Topology of Plane Sets of Points*. Cambridge University Press, Second Edition (1951).
‡Ore, O. *Theory of Graphs*, American Mathematical Society Colloquium Publications, vol. 38 (1962).
‡Prasolov, V. V. *Elements of Combinatorial and Differential Topology*, Graduate Studies in Mathematics No. 74, American Mathematical Society (2006).
‡Prasolov, V. V. *Elements of Homology Theory*, Graduate Studies in Mathematics No. 81, American Mathematical Society (2007).
‡Ranicki, A. (ed.), *The Hauptvermutung Book: A Collection of Papers on the Topology of Manifolds*, Springer-Verlag (1996).
‡Ringel, G. *Map Color Theorem*, Springer-Verlag (1997).
†Sato, H. (trans. K. Hudson) *Topology: An Intuitive Approach*, Translations of Mathematical Monographs vol. 183, American Mathematical Society (1996, 1999).
†Seifert, H. and W. Threlfall, *Lehrbuch der Topologie*. Chelsea Publishing Company, (2006, reprint of the 1934 edition).
‡Spanier, E. H. *Algebraic Topology*, Springer-Verlag (1994).
†Tutte, W. T. *Graph Theory As I Have Known It*, Oxford University Press (1998).
†Wall, C. T. C. *A Geometric Introduction to Topology*, Dover Publications (1994).
†Wilson, R. J. *Introduction to Graph Theory*, Prentice-Hall (1996).
†Wilson, R. J. *Four Colours Suffice: How the Map Colour Problem was Solved*, Penguin Books (2003).
†Wilson, R. J. and L. W. Beinecke (eds.), *Applications of Graph Theory*, Academic Press (1979).

Articles in journals and books

Appel, K. A. & W. H. Haken, 'The solution of the four-color-map problem', *Scientific American*, vol. 237, Oct. 1977, 108–121.
Bernstein, I. N., I. M. Gel'fand & V. A. Ponomarev, 'Coxeter Functors and Gabriel's Theorem', *Russian Mathematical Surveys*, vol. 28 (1973), 17–32.
Bing, R. H. 'Some aspects of the topology of 3-manifolds related to the Poincaré conjecture'. In *Lectures on Modern Mathematics* Vol. II, ed. T. L. Saaty, Wiley (1964), pp. 93–128.

References

Bryant, V. 'Straight line representations of planar graphs', *Elemente der Mathematik*, vol. 44 (1989), 64–65.
Doyle, P. H. & D. A. Moran, 'A short proof that compact 2-manifolds can be triangulated', *Inventiones Mathematicae*, vol. 5 (1968), 160–162.
Fáry, I. 'On straight line representations of planar graphs', *Acta Scientiacum Mathematicarum Szeged*, vol. 11 (1948), 229–233.
Franklin, P. 'A six colour problem', *Journal of Mathematics and Physics*, vol. 13 (1934), 363–369.
Freedman, M.H. 'The topology of 4-dimensional manifolds', *Journal of Differential Geometry*, vol. 17 (1982), 357–453.
Gardner, M. 'Of sprouts and Brussels sprouts, games with a topological flavor', *Scientific American*, vol. 217, July 1967, 108–121; and Chapter 1 in Gardner, M. *Mathematical Carnival*, Penguin (1990). [All Martin Gardner's columns from *Scientific American* are available on a CD, published by the Mathematical Association of America, 2005.]
Gardner, M. 'On the remarkable Császár polyhedron and its applications in problem solving', *Scientific American*, vol. 232, May 1975, 102–107; and Chapter 11 in Gardner, M. *Time Travel and Other Mathematical Bewilderments*, W.H. Freeman (1987).
Gardner, M. 'In which a mathematical aesthetic is applied to modern minimal art', *Scientific American*, vol. 239, Nov. 1978, 22–32, and Chapter 8 in Gardner, M. *Fractal Music, Hypercards and More*, W.H. Freeman (1991).
Goldstein, R. Z. & E. C.Turner, 'A formula for Stiefel-Whitney homology classes', *Proceedings of the American Mathematical Society*, vol. 58 (1976), 339–342.
Halperin, S. & D. Toledo, 'Stiefel-Whitney homology classes', *Annals of Mathematics*, vol. 96 (1972), 511–525.
Heawood, P. J. 'Map colour theorem', *Quarterly Journal of Mathematics*, vol. 24 (1890), 332–338.
Hilton, P. J. 'A brief, subjective history of homology and homotopy theory in this century', *Mathematics Magazine*, 60 (1988), 282–291.
Jungerman, M. & G. Ringel, 'Minimal triangulations of orientable surfaces', *Acta Mathematica*, vol. 145 (1980), 121–154.
Mac Lane, S. 'A combinatorial condition for planar graphs', *Fundamenta Mathematicae*, vol. 28 (1937), 22–32.
Milnor, J. W. 'Two complexes which are homeomorphic but combinatorially distinct', *Annals of Mathematics*, vol. 74 (1961), 575–590.
Poincaré, H. 'Cinquème complément á l'analysis situs', *Rendiconti del Circolo Matematico di Palermo*, vol. 18 (1904), 45–110.
Puppe, D. 'Korrespondenzen in abelschen Kategorien', *Mathematische Annalen*, vol. 148 (1962), 1–30.
Ringel, G. 'Wie man die geschlossenen nichtorientierbaren Flächen in möglichst wenig Dreiecke zerlegen kann', *Mathematische Annalen*, vol. 130 (1955), 317–326.
Ringel, G. 'Another proof of the map color theorem for non-orientable surfaces', *Journal of Combinatorial Theory, Series B*, vol. 86 (2002), 221–253.
Ringel, G. & J. W. T. Youngs, 'Solution of the Heawood Conjecture', *Proceedings of the National Academy of Sciences U.S.A.*, vol. 60 (1968), 438–445.
Schafer, J. A. 'Representing homology classes on surfaces', *Canadian Mathematical Bulletin*, vol. 19 (1976), 373–374.

Thomassen, C. 'Kuratowksi's theorem', *Journal of Graph Theory* vol. 5 (1981), 225–241.

Wagner, K. 'Bemerkungen zum Vierfarbenproblem', *Jahresbericht der Deutschen Mathematiker-Vereinigung*, vol. 46 (1936), 26–32.

Whitney, H. 'Non-separable and planar graphs', *Transactions of the American Mathematical Society*, vol. 34 (1932), 339–362.

Index

Not all phrases containing more than one keyword are indexed under each keyword. Thus "abstract graph" appears only under "graph", "homology group" only under "homology", etc.

abelian group, 215
  cyclic, 216
  direct sum, 222
  finitely generated, 217
  free, 218, 219
  generator of subgroup, 216
  internal direct sum, 223
  invariant factors, 231
  order of, 216
  order of element, 216
  presentation of, 227
  quotient, 220
  rank of, 234
  Smith normal form, 231
  subgroup, 216
  torsion coefficients, 231
  trivial, 216
abelian groups
  classification theorem for f.g., 231
adjacent, 27, 30, 209
affine
  dependence, 70
  independence, 68, 70, 72, 78, 86, 145
  span, 70
  subspace, 69, 95
Ahlfors, 133
Ahlfors, L.V., 38
alternating sum of dimensions, 177
alternating sum of ranks, 236
arc, 9, 89
  polygonal, 26
  simple polygonal, 26
augmentation, 25, 103, 106, 140
  mod 2, 171

augmented chain complex, 103
  mod 2, 173

ball, 196
  non-collapsing, 93
  three-, 92
barycentre, 136
barycentric
  coordinates, 68, 70
  subdivision, 48, 49, 55, 128, 136, 183
base of cone, 145
basis
  continuous family, 94
  for 1-cycles, 23, 29, 30, 105
  for 1-cycles mod 2, 175
  for direct sum, 223
  for free group, 218, 234
  for homology of surface, 152, 154, 176
  orientation, 93
  ordered, 93
  preferred, 93
  standard, 218
Beinecke, L.W., 9
Berge, C., 9
Bernstein, I. N., 25
Betti numbers, 155
Bing, R. H., 93
Birkhoff, G., 215, 231, 233, 234
Bollobás, B., 9
boundary, 104, 124
  edge, 38
  homomorphism, 21, 25, 101, 103
  homomorphism mod 2, 172
  mod 2, 174

243

boundary ctd.
  of chain, 101
  of edge, 22
  of simplex, 70, 75, 100, 148
  of surface, 187, 188
  relative, 113, 115
  relative homomorphism, 113
  relative homomorphism mod 2, 176
bounded subset of plane, 27
Brussels sprouts, 211

category, 234, 237
Cayley's theorem, 17
cell complex, 83
chain, 20, 99
  associated to path, 21, 22
  boundary of, 101
  complex, 103, 114, 124
  complex mod 2, 173
  group, 21, 99
  group of K mod L, 113
  map, 124
  mod 2, 171
  relative, 113
  relative group, 113
  with coefficients in A, 173
  with one simplex, 100
characteristic classes, 178
chromatic number, 65, 211
circle, 7, 8, 133
  chords, 29
closure, 84
coboundary homomorphism, 36
cochain, 36, 191
coherent orientation, 44
Cohn-Vossen, S., 42
cohomology, 191, 194
collapse, 19, 89, 144, 185, 189
  elementary, 19, 89
  invariance under, 144, 176
  of disk, 92
  to a point, 89, 145, 146
colourable, 65
colouring problem, 65, 211
component
  of graph, 17, 27
  of regular neighbourhood, 183
  of simplicial complex, 77, 106
  path, 27, 98
cone, 86, 145–147, 162, 188
  homology mod 2 of, 176

homology of, 145
connected, 40
  graph, 17
  simplicial complex, 77
  sum, 39, 54, 60, 165, 178
connectivity number, 174
continuous family
  of bases, 94
  of matrices, 96
convention, 75, 100, 115
convex, 72, 76
  hull, 72
Conway, J. H., 211
coordinates, barycentric, 68, 70
coset, 220
Courant, R., 27
cross-cap, 51, 52
cubic, twisted, 11
current, linear equations for, 15, 35
curve
  rational normal, 78
curved line, 31
cycle, 1, 21, 104, 105, 124
  fundamental, 112
  mod 2, 173, 176
  relative, 114, 115
  representative, 105, 114
cyclic group, 216
cyclomatic number, 14, 16, 19, 26, 87
cylinder, 38, 43, 82, 92, 122, 148, 188, 189
  genus of, 188
  homology of, 110, 118, 119, 141, 145
  Klein bottle becomes, 204
  regular neighbourhood is, 203
  two-sided, 196
  union of, 163

Damphousse, P., 31
dancer, Cossack, 13
dependence, affine, 70
diagram
  chasing, 169
  commutative, 124, 169
Diestel, R., 9
differential graded group, 126
dimension
  of abstract simplicial complex, 78
  of affine subspace, 69
  of simplex, 70
  of simplicial complex, 74, 75
direct sum, 126, 222, 223, 238

Index

disk
  collapsing of, 91
  Euler characteristic of, 92
  region a, 194, 208, 209
  region not a, 208
  triangulation of, 134
  two-dimensional, 38, 56, 89, 200
domain
  of homomorphism, 217
double of surface, 189
drawing
  graphs, 12
  simplicial complexes, 75, 79
dunce hat, 82, 145

edge
  boundary, 38
  boundary of, 21
  inner, 18
  of abstract graph, 9
  of abstract oriented graph, 10
  of curved graph, 32
  of graph, 10
  of oriented graph, 11
  removing an, 27, 209
Eilenberg, S., 5, 7, 221
electrical circuit, 10, 14
electro-motive force, 15, 35
embedding of graph in surface, 62, 180
epic, 217
epimorphism, 217
equivalent
  closed surfaces, 53, 132
  p.l., 132, 134, 209
Euler characteristic, 55, 91, 154, 177
  invariance under subdivision, 60
  of closed surface, 56, 64
Euler's formula, 28, 29, 56, 181
  for curved graphs, 32
  generalization of, 207
Euler's inequality, 207
exact sequence, 139, 140, 221
  split, 223
  short, 141, 164, 221, 234
excision theorem, 142
  mod 2, 176

face, 72
  free, 88, 92
  proper, 72

Fáry, I., 31, 32, 50, 59, 134, 182, 200, 208
finitely generated, 104, 217
five colour theorem, 66, 211
forest, 26
four colour theorem, 66
frame, 93
Franklin, P., 65
free
  abelian group, 218, 219, 224, 226
  generators, 218
  vertex, 18
Freedman, M., 134
frontier of region, 30
functor, 234, 237
fundamental
  class, 112
  cycle, 112, 149
  cycle mod 2, 174

Gardner, M., 81, 211
Gel'fand, I. M., 25
general linear group, 98
general position, 205
generator(s)
  changing, 227
  for direct sum, 223
  free, 218
  of cycle group, 21, 24, 105
  of cyclic group, 216
  of homology group, 106, 107, 153, 154, 163
  of relative cycle groups, 119
  of relative homology group, 116, 140
  of subgroup, 216
genus, 53, 188
Goldstein, R.Z., 179
graph, 11, 75
  n-colourable, 65
  abstract, 9
  abstract oriented, 10
  bipartite, 24, 31, 87
  complete, 10, 31, 62, 64, 65, 73, 208
  component of, 17
  conditions for planarity, 30
  connected, 17
  curved, 32
  cyclomatic number of, 14, 16, 19, 20, 26, 87
  drawing, 12
  in sphere, 182
  in surface, 62, 180
  in torus, 28, 62, 120, 181, 198, 208

graph ctd.
  isomorphism of, 13
  no isolated vertices, 39
  non-separating, 199, 200
  oriented, 11
  planar, 26, 182
  simple, 9
  subdivision of, 31, 208
  suspension of, 162
  triangular, 32, 210
Gross, J., 9

Halperin, S., 179
handle, 52, 63
Hauptvermutung, 134
Heawood conjecture, 65
Heawood theorem, 65, 211
Heawood, P. J., 65
Hilbert, D., 42
Hilton, P. J., 126, 133, 179
hole(s)
  p-dimensional, 99, 104
  independent, 105
  punching, 200
homeomorphism, 6, 127, 133
  of plane, 32
homologous, 105
  mod L, 114
  mod 2, 174
  to zero, 3, 105
homology class, 105
  mod 2, 202
  relative, 114
  represented by simple closed polygon, 200
homology group(s), 104, 124
  0-th, 106
  0-th mod 2, 175
  0-th relative, 116
  1st, 3, 107
  1st mod 2, 175
  1st relative, 117
  2nd, 3, 111
  Alexander–Spanier, 7
  Čech, 7
  independent of orientation, 105
  invariance of relative, 132
  invariance under collapsing, 144, 162, 176
  invariance under stellar subdivision, 132
  mod 2, 174
  mod 2 of cone, 176
  mod 2 of graph, 175
  mod 2 of Möbius band, 175
  mod 2 of Möbius band mod rim, 177
  mod 2 of projective plane, 179
  mod 2 of n-sphere, 177
  mod 2 of surfaces, 174, 176
  of n-sphere, 148, 157, 162
  of 2-simplex mod boundary, 115
  of closed surface, 111, 152
  of cone, 146
  of cylinder, 110, 141, 145
  of cylinder mod ends, 118, 141
  of cylinder mod one end, 120, 141
  of graph, 104, 142
  of Klein bottle, 153, 163, 165, 166
  of Möbius band, 108, 141, 145
  of Möbius band modrium, 119, 141
  of projective plane, 111, 142
  of simplex, 146
  of simplex mod boundary, 115
  of surface, 111, 152, 189, 190
  of torus, 111, 163
  of union, 107, 161
  reduced, 104, 140
  reduced, mod 2, 174
  relative, 4, 114, 140, 146
  relative mod 2, 176
  singular, 7
  top dimensional, 105
  topological invariance, 6, 127, 136
homology sequence
  of a pair, 139
  of a triple, 168
  mod 2 of a pair, 176
  mod 2 of a triple, 177
homomorphism, 216
  splitting, 223
  theorem, 220
  trivial, 217
homomorphism theorem, 220

image, 217
independence, affine, 68, 70, 72, 78, 86
indivisible, 202, 237
inside, 27, 196
interior of simplex, 71
intersection condition
  for closed surfaces, 39
  for graphs, 11

for simplicial complexes, 74
violated, 48, 76, 83
invariance
   topological, 6, 127, 136
   under stellar subdivision, 132
isomorphic
   abstract graphs, 13
   groups, 217
   abstract simplicial complexes, 78
   graphs, 13
   simplicial complexes, 78
isomorphism, 217

Join
   of simplexes, 86
   of simplicial complexes, 86, 148, 156
joinable, 86
Jordan curve theorem, 32–34
   for polygons, 27, 181
Jungerman, M., 62

Kelley, J. L., 127
kernel, 217
kinky, 39
Kirchhoff's laws, 14, 19, 20, 23, 35
Klein bottle, 4, 42, 44, 47, 54, 55, 64, 121, 153, 163, 173, 205
   becomes cylinder, 204
   Brussels sprouts on, 214
   chromatic number, 65
   homology of, 153, 163, 165, 166
   homology mod 2 of, 178
   minimal triangulation of, 61
   one-sided?, 196
   realization in $\mathbb{R}^4$, 58
Kuratowski, K., 30

Lagrange's theorem, 220
Lang, S., 215, 231, 233
Lefschetz duality, 194
letter, 49
linear
   combination, 218
   map, 173, 224, 234
link, 40, 87, 187
oop(s) (*see also* simple closed polygon)
   1-chain associated to, 22
   basic, 20, 23, 35, 152
   graph with one, 20
   independent, 15, 24

on graph, 15, 17
orientation preserving, 203
orientation reversing, 203
oriented, 25

Mac Lane, S., 30, 215, 231, 233, 234
manifold
   3-dimensional, 197
   combinatorial, 182
Maxwell, J. C., 16
Mayer–Vietoris sequence, 159
   mod 2, 178
   reduced, 161
Milnor, J. W., 5, 134, 178
minimal triangulation, 60
   of Klein bottle, 61
   of non-orientable surface, 62
   of orientable surface, 62
   of projective plane, 61
   of sphere, 61
   of torus, 61, 81
Möbius band, 38, 43, 44, 58, 82, 92, 122, 147, 165, 173, 188, 189
   double, 58
   genus of, 188
   homology of, 108, 119, 141, 145
   homology mod 2 of, 175, 177
   one-sided, 196
   regular neighbourhood a, 203
   union of, 165
monic, 217
monomorphism, 217
Morton, H. R., 203

natural projection, 220
network, 9
non-orientable (*see also* surface)
   closed surface, 44
   surface, 187
non-separating graph, 199, 200

octahedron, 6, 148
one-sided, 196
order
   of group/element, 216
   of vertex, 14, 33
ordered simplex, 73
orientable (*see also* surface)
   closed surface, 44, 47
   surface, 187

orientation
  class, 94
  coherent, 44, 149
  cycle group independent of, 24
  homology, independent of, 105
  of abstract graph, 10
  of abstract simplicial complex, 79
  of basis, 93
  of edge, 2, 11
  of graph, 11
  of simplex, 74, 95
  of simplicial complex, 77
  of triangle, 44, 74
  opposite, 94, 95
  preserving, 94, 203
  reversing, 94, 203
  same, 94, 95
outside, 27, 196

pair
  of first kind, 50
  of second kind, 50
parallel, 69
path
  components, 27, 98
  on graph, 16
  polygonal, 98
  simple, 17
pentagon, 34
permutation
  of basis vectors, 95
  of vertices of simplex, 74, 95,
    96, 100
piecewise linearly equivalent, 132,
  134, 209
Poincaré, H., 203
point, 67
  collapse to a, 89, 145, 146
polygonal
  arc, 26, 187
  representation, 39, 45, 92, 151, 154
polyhedron, 79
Ponomarev, V, A., 25
Prasolov, V.V., 38, 126, 133, 179, 182, 194
presentation, 227
  of homology group, 108
pretzel, 2 (*see also* torus, double)
projection
  natural, 220
projective plane, 39, 42, 44, 51, 53, 59, 122,
    153, 196

n-fold, 54
  homology of, 111, 142
  homology mod 2 of, 179
  in 3-manifold, 196, 198
  minimal triangulation of, 61
  minus 2-simplex, 165
  non-orientable regions on, 207
  one-sided, 197
  realization in $\mathbb{R}^4$, 58
  two-sided, 198
projective space, 196
pure simplicial complex, 40, 75, 87

quotient group, 220, 226

Ranicki, A., 134
rank, 231, 234
  alternating sum of, 236
  of subgroup, 236
  plus nullity theorem, 177
realization
  of abstract graph, 11
  of abstract simplicial complex, 78, 81
  of surface, 39, 57, 196
reduced
  homology group, 104, 140, 174
  Mayor-Vietoris sequence, 161
  mod 2 sequence of pair, 176
  sequence of pair, 140
region(s)
  a disk, 194, 208, 209
  all disks, 208, 209
  all triangular, 210
  determined by V_i, 184, 193
  edge adjacent to, 27, 209
  frontier of, 30, 210
  graph in Klein bottle, 121
  graph in plane, 27, 28
  graph in surface, 4, 180
  graph in torus, 28, 120, 181, 208
  inside circle, 29
  minimal number of, 208
  non-orientable, 207
  not a disk, 208
  number of, 184, 194
  orientable, 206
  surface in 3-sphere, 195
regular neighbourhood, 183, 184
  a cylinder, 203
  a Möbius band, 203

collapsing, 185
relations, 227
  changing, 227
relative
  boundary, 114, 115
  boundary mod 2, 176
  chain, 113
  cycle, 114, 115
  homology class, 114
  homology group, 4, 114, 116, 132
  homology mod 2, 176
resistance, 15, 36, 37
restriction homomorphism, 116, 125, 139, 160
rim of Möbius band, 109, 119, 141, 147, 177
Ringel, G., 62, 63, 211
Ringel–Youngs theorem, 64, 65
Robbins, H. E., 27

Sario, L., 38, 133
schema, 47, 50, 63, 79
scissors and glue, 39, 49
segment, 11, 67, 71, 86
self-intersection, 42, 51, 58
separation
  of $\mathbb{R}^3$ by surface, 195, 196
  of 3-manifold by surface, 197
  of plane, 26, 27, 32
  of surface by graph, 3, 198
simple closed polygon, 27, 35, 40, 46, 87, 89, 91, 92, 145, 187
  boundary component of surface, 188, 204
  in surface, 198, 200
  in torus, 200, 205
  intersecting, 205
  orientation preserving, 203
  orientation reversing, 203
simplex(es), 70
  0- (*see also* vertex), 67, 71, 75
  1- (*see also* edge), 67, 71, 75
  2- (*see also* triangle), 71, 73, 75, 100
  boundary of, 71
  dimension of, 71
  face of, 72
  homology of, 146
  interior of, 71
  join of, 86
  join of oriented, 87
  of abstract simplicial complex, 78
  open, 71
  ordered, 73

oriented, 74, 95, 96
principal, 89
skeleton of, 156
simplicial approximation, 135
simplicial approximation theorem, 135
simplicial complex(es), 6, 40, 74
  abstract, 78, 80
  connected, 77
  dimension of, 74
  drawing, 75
  empty, 75, 86, 100, 147
  Euler characteristic of, 91
  in $\mathbb{R}^3$, 195
  join of, 86, 148, 156
  orientation of, 79
  oriented, 77
  p.l. equivalent, 132, 134, 209
  pure, 40, 75, 87
  realization of, 78
  skeleton of, 77, 115, 142
  spine of, 92, 189
  subdivision of, 128, 132
  underlying space of, 74, 79, 127
simplicial map, 130
skeleton, 77, 115
  1-, 77, 142
  homology groups of, 105, 156
  of simplex, 156
Smith normal form, 231
Spanier, E. H., 7, 178, 194
spanned, 70
sphere, 2, 39, 50, 53, 58, 92, 148
  n-, 148, 157, 162, 177
  3-, 195
  chromatic number of, 65
  minimal triangulation of, 61
  triangulation of, 6, 134
  with cross-cap, 51 (*see also* projective plane)
  with handle (*see also* torus), 52
spine, 92, 189
splitting lemma, 223
sprouts, 211
  Brussels, 211
standard form of closed surface, 49, 53
  calculation of, 57
  uniqueness of, 56
star, 85, 135, 200
  closed, 85, 131
starring, 88, 128
Stasheff, J. D., 178

Steenrod, N., 5, 7, 221
stellar subdivision, 88, 128
  invariance under, 132
Stiefel–Whitney classes, 178
subcomplex(es), 76, 77, 84
  disjoint, 107, 161
  intersection of, 77, 158
  proper, 77
  union of, 77, 158, 161
subdivision
  barycentric, 48, 49, 55, 83, 128, 136, 184
  of closed surface, 60, 209
  of graph, 31, 208
  of simplicial complex, 128, 132
  stellar, 88, 128
subgraph, 18
subgroup, 216
  cyclic, 216
  torsion, 111, 232, 237
sum
  connected, 39, 54, 60, 165, 178
  direct, 126, 222, 223
surface(s), 38, 187
  adding cones to, 188, 203
  boundary of, 187, 188
  Brussels sprouts on, 211
  chromatic number of, 65, 66
  closed, 39, 75, 87, 144, 162
  connected sum of, 39, 54, 60, 165, 178
  curved, 5, 38, 62, 133, 211
  cutting along loop, 204
  decisive property of, 38
  double of, 189
  divided into triangles, 210
  equivalent, 53, 132
  generators for homology, 153, 154
  genus of, 53, 188
  graph in, 62, 180
  homology of, 111, 152, 190
  homology mod 2 of, 174, 176
  in $\mathbb{R}^3$ or $\mathbb{R}^4$, 57, 195
  in 3-manifold, 197
  non-orientable regions on, 207
  one-sided, 196
  orientability, 43, 47, 187
  orientable regions on, 206
  p.l. equivalent, 132, 134, 209
  polygonal representation of, 45, 92, 151
  realization of, 47
  removing 2-simplex from, 150, 162, 165, 175, 176, 178, 188

separation by graph, 4, 198
standard form of, 49, 53, 56, 57
two-sided, 196
suspension, 162
symbol(s), 46, 49, 55
  juxtaposition of, 53

target
  of homomorphism, 217
tensor product, 179, 233
tetrahedron
  hollow, 6, 40, 45, 47, 133, 148
  solid, 67, 71
Toledo, B., 179
topological invariance, 6, 127, 136
torsion
  coefficients, 231
  product, 179
  subgroup, 111, 232, 238
torus, 2, 39, 41, 44, 52, 53, 79, 102, 163, 166, 189
  n-fold, 54, 154
  as cell complex, 83
  double, 2, 4, 54, 82, 154, 167, 199
  graph in, 28, 62, 120, 181, 199, 208
  homology of, 111, 163
  homology mod 2 of, 178
  in 3-manifold, 197
  knotted, 3
  minimal triangulation of, 61, 81
  obtained from plane, 202
  one-sided, 197
  simple closed polygons on, 200, 205
  solid, 3, 197
tree(s), 17
  Christmas, 13, 17
  maximal, 18, 19, 24, 142, 148, 175
  spanning, 18
  union of, 26, 210
triad, 159
triangle (see also simplex, 2-), 39, 67, 68, 71, 148
  flat, 39
  orientation of, 44, 74
  removing, 143, 150, 162, 165, 175–178, 188
  spherical, 58
triangulation, 5, 79, 133
  minimal, 60
  of surface, 38
triple, 167, 177

Turner, E.C., 179
two-sided, 196

unbounded
  region, 27
  subset of plane, 27
underlying space, 74, 79, 86, 127
universal coefficients theorem, 179

Vandermonde determinant, 12, 78
vector, 67
vertex
  free, 18
  Kirchhoff equation for, 15, 35
  of abstract graph, 9
  of abstract oriented graph, 10
  of abstract simplicial complex, 78
  of curved graph, 32
  of graph, 11
  of simplex, 72
  order of, 14, 33
  terminal, 18

Wall, C. T. C., 32
Whitney homology classes, 178
Whitney, H., 30
Wilson, R. J., 9, 66, 211
Wylie, S., 126, 133, 179

Yellen, J., 9
Youngs, J. W. T., 63

Printed in the United States
By Bookmasters